THE AI FUTURE

How to keep up and stay ahead

Peter Thorpe

vol 2 2026

Disclaimer:
The information contained in this book is intended for general informational purposes only. The book delves into various aspects of Artificial Intelligence (AI), providing insights, explanations, and examples to educate readers about this fascinating field.

It is important to note that certain sections of this book utilize content generated by AI language models, including but not limited to OpenAI's GPT-3. Every attempt has been made to ensure that this AI-generated content is original, accurate, and relevant to the topic at hand. The AI-generated content is meant to augment and enhance the overall understanding of AI concepts and applications.

However, as with any AI-generated content, there may be occasional instances where similarities or resemblances to existing works or sources could arise inadvertently. The author and publisher wish to emphasize that no intentional plagiarism or usage of copyrighted material from third parties has occurred in the production of this book.

Readers are encouraged to use the content within this book responsibly and in accordance with applicable laws and regulations. The information presented here is not intended as legal, technical, or professional advice, and any decisions or actions based on the content of this book are solely the responsibility of the reader.

The author and publisher do not make any warranties or representations concerning the accuracy, completeness, or reliability of the information presented herein. The content may be subject to change and should not be considered an exhaustive or definitive source on AI-related matters.

By reading this book, the reader acknowledges and accepts that the use of AI-generated content, while aimed at providing valuable insights, is an evolving technology and may not be entirely devoid of unintended similarities to existing works.

IMPORTANT NOTE
Artificial Intelligence is evolving at an extraordinary pace. Apps, features, menus and capabilities are frequently updated—sometimes weekly—so, some descriptions or functions may appear slightly different to how they are described in this book.

I try to update the book regularly, usually at least once a month, to attempt to reflect the most significant changes, but it's impossible to stay perfectly in sync with every new release across various products. So, if something appears slightly different to what's mentioned in the book, don't be concerned. Most changes are made to improve your experience, and the core ideas and techniques usually remain the same. If you're ever unsure, just ask an AI like ChatGPT directly for help.

Peter Thorpe
Editor

TABLE OF CONTENTS

THE INTELLIGENCE REVOLUTION

Something extraordinary is unfolding in our world — not quietly, but at breathtaking speed. We are living through the early stages of what may come to be known as *the Intelligence Revolution*. Like the *Industrial Revolution,* that mechanized human muscle, this one is amplifying human thought itself. For the first time in history, the power to reason, imagine, and create is being shared between billions of people and the intelligent systems they now carry in their pockets.

This isn't a story about some distant future. It's happening right now. Across the world, millions of people open an app like ChatGPT, Gemini or Claude every single day and ask it to write, analyse, summarise, design, explain, or even teach. These aren't researchers or coders anymore — they're students, small-business owners, retirees, and everyday workers using AI as naturally as they once used Google. What began as a curiosity has become a companion.

THE TIPPING POINT
Every great technological shift has its tipping point — the moment when something that once felt experimental becomes normal. For artificial intelligence, that moment has already arrived. When more than a billion people start using conversational AI tools regularly, you no longer have a specialist technology — you have an infrastructure for everyday thinking.

And what's driving it isn't just hype or novelty; it's economics and accessibility. Only a few years ago, using a leading AI model cost serious money. Running complex tasks was the domain of companies, universities, or well-funded labs. But advances in computing efficiency and the falling cost of training and serving AI models have rewritten the rules. Processing power that once cost tens of dollars per million words can now be delivered for cents.

That's why we now find AI built into word processors, search engines, design tools, and even email. Like electricity or the internet, it's becoming invisible — just another layer of daily life.

FROM COMPLICATED TO CONVERSATIONAL
Until recently, using AI well required a kind of secret handshake — careful prompts, specific phrasing, and plenty of trial and error. But that barrier has been swept away by design. The new generation of AI systems speaks *our* language, not the other way around. You can describe an idea as if you were explaining it to a friend — *"make me a logo that feels friendly but professional,"* or *"turn this photo into a 1950s poster"* — and the system simply does it.

That ease of use has unlocked a wave of creativity. Writers are experimenting with plot ideas. Farmers are generating marketing flyers. Builders are designing extensions or drafting job quotes. Teachers are preparing lesson plans, and health workers are simplifying patient notes. Each example is small on its own, but together they mark a profound shift: the moment when the average person gains access to world-class cognitive tools without needing special training.

WHEN INTELLIGENCE BECOMES ABUNDANT
For most of human history, intelligence has been rare, expensive, and slow to reproduce. Societies were organised around this scarcity — schools to transmit knowledge, professions to guard expertise, and universities to certify who possessed it. But now we are entering an age where intelligence, or at least a form of it, is abundant.

When you can ask a machine to draft a legal contract, translate a document, write a business plan, or tutor a child — instantly and at negligible cost — the foundations of education, work, and creativity all start to shift.

Welcome to the Intelligence Revolution!

The implications are enormous. Small businesses can compete globally. A student in a remote town can access the same level of analysis as someone in a major university. An individual with an idea can build a prototype, market it, and find customers — all with digital assistance. In many ways, AI is levelling the playing field.

But it also introduces a new imbalance — between those who *use* it effectively and those who don't. Knowing how to collaborate with AI may become as essential as literacy once was. The revolution isn't about replacing human intelligence; it's about multiplying it.

THE PARADOX OF PROGRESS
Of course, abundance always brings complexity. For every uplifting use of AI, there's a shadow side. The same systems that help doctors detect diseases, can generate convincing fake videos. The same chatbots that offer companionship, can distort reality or encourage unhealthy dependence. The same efficiency that empowers workers can also displace them.

In that sense, we are entering what you might call *the messy middle* of the Intelligence Revolution — a period of enormous creativity, confusion, and contradiction. Our tools are astonishing, but our ability to manage their consequences is still catching up.

Think of it this way: society is built on trust — in information, in institutions, in one another. But AI is blurring those boundaries. When anyone can produce a realistic photo, voice, or video in seconds, how do we know what's authentic? When machines can pass exams or write essays, how do we measure genuine understanding? These questions don't have easy answers yet, but they're forcing every field — education, media, law, and governance — to rethink its assumptions.

THE NEW PARTNERSHIP
Still, there's another, more optimistic way to view this revolution: as a partnership between human creativity and

machine capability. AI is not conscious, not emotional, and not wise — but it is extraordinarily capable at amplifying what we feed it. In skilled hands, it becomes a multiplier of imagination.

The most successful people in this new era will be those who learn to ask better questions, frame problems better, and use AI to explore possibilities faster than ever before. *Prompting, reviewing,* and *refining* are becoming the new creative disciplines. It's less about programming machines and more about collaborating with them.

Already we're seeing early glimpses of that partnership: scientists using AI to discover new materials, filmmakers generating storyboards in hours instead of weeks, not-for-profits translating health information into dozens of languages overnight. What was once slow and costly, now happens at the speed of thought.

ADAPTING OUR INSTITUTIONS
Every major technological revolution eventually forces society to reinvent itself. The steam engine transformed cities and labour. Electricity redefined industry and home life. The internet reshaped communication and commerce. Now AI is challenging our assumptions about knowledge, expertise, and even creativity itself.

Schools will have to shift from teaching memorisation to cultivating judgment. Workplaces will focus less on output and more on oversight and design. Governments will wrestle with how to regulate systems that evolve faster than legislation. The change will not be smooth, but it will be inevitable.

What makes this revolution distinct is its reach. Unlike past technologies that affected specific industries first, AI touches *every* discipline that relies on information — which is to say, almost all of them. It doesn't just change what we can do; it changes how we think about doing it.

HUMAN INTELLIGENCE IN AN AI WORLD

And that brings us to the deepest question of all: where do we, as humans, fit in when intelligence is no longer scarce? If machines can simulate reasoning, produce art, and offer advice, what remains uniquely ours?

The answer may lie not in competing with AI but in combining with it. *Empathy, ethics, humour, intuition* — these are not things a machine can truly master. Our role is to bring the human layer: *context, compassion,* and *moral insight.* The more intelligent our tools become, the more valuable those human qualities will be.

We are not being replaced; we are being expanded. *The Intelligence Revolution* isn't about handing over control to machines — it's about learning to live with them and think alongside them. It's about designing a future where artificial intelligence enhances our natural intelligence, and where progress serves people, rather than the other way around.

THE ROAD AHEAD

So yes, the world is getting faster, stranger and more complex. But it also presents more possibilities than ever before. Just as the industrial age built the physical world we inhabit; the intelligence revolution is building the cognitive world we'll navigate next.

We can treat it with fear or with curiosity. We can fight it, or we can shape it. The choice, as always, is ours. Because this revolution — the Intelligence Revolution — is not about the rise of machines. It's about the next great leap in what it means to be human.

HYPE TO REALITY:
WHAT AI CAN AND CAN'T DO TODAY

AI sits at the centre of one of the biggest hype cycles in history. On one hand, some claim it will solve climate change, cure every disease, and solve world peace. On the other hand, critics warn that AI could wipe out jobs, spread misinformation, or even threaten humanity's survival.

The truth is less dramatic but no less important. AI is an extraordinary tool — but it has clear limits and there are risks.

WHAT AI EXCELS AT

Language and Communication
Large language models like ChatGPT can generate human-like text — writing emails, reports, speeches, stories, and even computer code. They can translate between languages, summarise complex information, and assist with brainstorming.

For example, lawyers are using AI to draft contracts faster. Students are using it to structure essays. Businesses are using it to create marketing content in seconds.

Pattern Recognition
AI is particularly powerful at spotting patterns in large datasets — something humans struggle with. It can detect anomalies in financial transactions *(useful for fraud prevention),* identify diseases in medical scans, or forecast consumer trends based on massive sales data.

Automation of Repetitive Digital Tasks
Customer service chatbots, transcription services, scheduling assistants, and data-entry automation are already common. These free up humans to focus on higher-level or more creative work.

WHERE AI STRUGGLES

Common Sense and Context
AI doesn't *"understand"* the world the way humans do. It predicts the most likely next word or outcome based on patterns in its training data. This means it can produce text that *sounds* intelligent but is logically wrong or factually inaccurate — a phenomenon often called *"hallucination"*.

Physical Work
Robots exist, but they are not yet capable of replacing plumbers, electricians, aged carers, or mechanics. Human dexterity, adaptability, and problem-solving in the physical world remain unmatched.

Ethics, Values, and Judgment
AI can't decide what's right or wrong. It has no values — only probabilities. Decisions about fairness, justice, and responsibility, remain firmly human responsibilities.

CUTTING THROUGH THE HYPE
When you read headlines like *"AI will replace doctors"* or *"AI will end education, and we'll all become vegetables"* treat them with caution. The reality is that AI is a tool that can *assist* doctors and *support* teachers — but not replace them. In fact, in many cases AI may make human roles more valuable, not less.

Think of AI today as a **very powerful assistant**: it can draft, suggest, and automate. But the ultimate responsibility and creativity lies with you.

THE KEY TURNING POINTS

AI didn't suddenly appear in 2022 when ChatGPT went viral with public access. It has been building for decades, through a series of key turning points. Understanding these moments will help you see where AI might go next.

THE EARLY DREAMS (1950s–1980s)
The idea of *"thinking machines"* goes back to pioneers like Alan Turing, who in 1950 proposed the famous *"Turing Test"* to measure machine intelligence. Early experiments with *"neural networks"* attempted to mimic how the brain worked. But progress was painfully slow — computers were too weak, and data was too scarce. AI research went through *"winters"* when enthusiasm and funding dried up.

THE INTERNET AND BIG DATA (1990s–2000s)
The arrival of the internet was a game-changer. Suddenly, data was everywhere — text, images, video, and more. This gave researchers the raw material to train more powerful algorithms. The internet gave developers access to massive amounts of data and material and search engines like Google demonstrated the power of large-scale pattern recognition.

THE RISE OF DEEP LEARNING (2010s)
A major breakthrough came with *"deep learning"* a technique using many layers of artificial neural networks. In 2012, a deep learning system shocked the world by winning an image-recognition contest, correctly identifying cats, dogs, and objects from millions of photos. This success kicked off a decade of rapid progress.

Voice assistants like *Siri* and *Alexa*, facial recognition systems, and early self-driving car prototypes, all grew out of this era.

TRANSFORMERS: THE LANGUAGE REVOLUTION (2017)
The single biggest leap came in 2017 with a new architecture called the **transformer.** Unlike older systems, transformers could process words in relation to each other, not just one by one. This made it possible for AI to truly *"understand"* language structure.

Within a few years, this led directly to GPT (Generative Pretrained Transformer) models, Google's *BERT*, and many other systems. Suddenly, AI could write coherent paragraphs, translate smoothly, and hold realistic conversations.

GENERATIVE AI GOES MAINSTREAM (2022–PRESENT)
The launch of ChatGPT in late 2022, marked the tipping point. For the first time, ordinary people could interact with AI in natural language — and it felt surprisingly human. At the same time, tools like *DALL·E, MidJourney,* and *Stable Diffusion* allowed anyone to generate images from text prompts.

By 2023–2024, AI could also create music, design products, and even produce short films. The *"generative"* era had begun, and adoption spread at record speed. Within two months of its launch, ChatGPT had over 100 million users. It was the fastest take up of any technology in history.

WHY THESE TURNING POINTS MATTER
Each step built on the last. Without the internet, there'd be no data. Without deep learning, no image recognition. Without transformers, no ChatGPT. And without ChatGPT's viral success, AI might still be stuck in research labs.

The history shows that AI is not random — it follows a trajectory of breakthroughs.

The next turning points will include:

- **AI Agents** *that can plan, reason, and act on your behalf.*

- **Multimodal AI** *that seamlessly combines text, images, video, and voice.*

- **Embedded AI** *in everyday tools, from word processors to home appliances.*

We've only just stepped into this new era.

SUMMARY
By understanding the pace of change, cutting through the hype vs the reality, and recognising the key turning points that brought us here, you now have the foundation to further explore the AI future.

The important takeaway is this:

AI is not magic, and it's not science fiction. It is a powerful human-made tool — and the more you understand it, the more effectively you can use it to future-proof yourself in the years ahead.

WELCOME TO THE AI FUTURE

Artificial Intelligence is no longer a futuristic concept—it's woven into the fabric of our daily lives. From smartphones that anticipate our needs to algorithms that shape the way we shop, learn, and work, AI has become both a silent partner and a powerful driver of change.

The pace of this technological revolution is faster than anything we've seen before, and its impact is broader than the arrival of the motor car, the telephone, the personal computer, or even the internet.

For individuals, the challenge is clear:

How do we not only keep up, but also stay ahead in this AI-driven future? The answer begins with awareness and adaptability.

WHY THIS MATTERS

AI is not just automating routine tasks; it is reshaping industries, redefining skills, and even creating entirely new career paths. Jobs that rely on repetition and predictable patterns are already being transformed by AI systems. Administrative work, data entry, basic bookkeeping, and even some customer service roles are being taken over by automated assistants and smart software. At the same time, entirely new opportunities are emerging. In some cases, AI is replacing jobs, however, it's also creating jobs that didn't exist prior to AI.

JOBS THAT COULD BE REPLACED BY AI

Retail Cashiers
Check out people in retail stores

Drivers
Drivers of cars, trucks, buses and trains

Data Entry Clerks
Typing information into databases.

Bookkeepers
Keeping track of accounts and expenses.

Customer Service Representative
Handling customer questions or complaints.

Telemarketers
Calling people to sell items or services.

Translators
Translating between languages.

Factory Workers
Working on a production line doing repetitive tasks.

JOBS THAT HAVE BEEN CREATED BY AI

Here are just some of the jobs that have been created by AI:

AI Ethics and Governance
Ensuring AI is used responsibly.

Prompt Engineering
Designing the questions and instructions that guide AI systems.

Human–AI Collaboration
Creating workflows where people and AI work together and complement each other.

AI Product Manager
Bridging the gap between technical teams and business needs — deciding what AI features to build and how to deploy them.

AI Maintenance and Oversight
Training, testing, and improving AI models.

AI Trainer / Model Fine-Tuner
Teaching AI systems how to respond better to human input, often by reviewing and ranking model outputs.

AI Integration Specialist
Helping companies incorporate AI tools (like ChatGPT or Copilot) into their workflows.

AI Product Manager
Bridging the gap between technical teams and business needs — deciding what AI features to build and how to deploy them.

AI Security Analyst
Protecting AI systems from manipulation, bias attacks, and data breaches.

AI Legal Advisor / Policy Consultant
Advising businesses and governments on regulations, copyright, and ethical implications of AI use.

Human Empathy Coordinator / Customer-AI Liaison
Ensuring that automated customer interactions remain empathetic and human-centred — combining emotional intelligence with tech skills.

This transformation makes it essential for everyone—students, professionals, retirees, and business leaders alike—to understand the basics of AI and adapt to its growing presence.

THE MINDSET SHIFT

Keeping up with AI requires a shift in thinking. Instead of fearing replacement, it's important to look at AI as a tool that can augment human abilities. Those who thrive will be the ones who learn how to work *with* AI, leveraging its speed, accuracy, and pattern recognition, while bringing human creativity, empathy, and critical thinking to the table.

A simple analogy is the introduction of the calculator. When calculators first appeared, some feared they would destroy mathematics education. Instead, they freed people from tedious calculations and allowed them to focus on problem solving and higher-level reasoning. AI is the next-level calculator—one that can process not just numbers, but languages, images, and even ideas.

The question isn't *"Will AI take my job?"* but rather, *"How can I use AI to make my work better, faster, or more impactful?"*

This quote from Jensen Huang, CEO of Nvidia, nails it:

"Instead of thinking about AI as replacing the work of 50% of the people, you should think that AI will do 50% of the work for 100% of the people. AI will not take your job. AI used by somebody else will take your job".

You need to constantly adapt, learn, and use AI, or risk being left behind.

PRACTICAL STEPS TO STAY AHEAD

Learn the Basics of AI Tools
You don't need to be a computer scientist, but understanding how tools like ChatGPT, image generators, and AI-powered search engines work will give you an edge. Start with free versions, watch tutorials, and try them out for everyday tasks.

Stay Curious and Keep Learning
The AI landscape is evolving rapidly. Treat learning as a lifelong habit. Read articles, follow trusted newsletters, or take short online courses. The more you expose yourself to AI developments, the less intimidating they will feel.

Adapt Your Skills
Focus on skills that are harder for AI to replicate: problem solving, communication, leadership, emotional intelligence, and hands-on trades. A plumber, aged-care worker, or counsellor, for example, offers human presence and trust—things AI cannot fully replace.

Experiment
Try using AI in your daily work or hobbies. Write an email draft with ChatGPT, create an AI-generated picture for a presentation, or use an AI scheduling assistant. The best way to understand it, is to interact with it directly.

Think Ahead
Consider where AI might change your industry or community and prepare early. Teachers might explore AI tutoring tools, healthcare workers might learn about AI diagnostics, and small business owners could test AI for marketing or customer service.

REAL-WORLD CASE STUDIES

Teachers
Many educators are beginning to use AI as a teaching assistant. For example, a high school teacher might ask ChatGPT to generate quiz questions, provide alternative explanations for struggling students, or design lesson plans tailored to different learning levels. This doesn't replace the teacher's role but frees up more time for one-on-one interaction with students.

Healthcare Workers
Doctors and nurses are already seeing the benefits of AI-

powered diagnostics. For instance, AI systems can scan medical images (like X-rays or MRIs) with remarkable speed and accuracy, flagging potential issues for a doctor to review. This allows healthcare professionals to focus more on patient care, rather than endless paperwork.

Small Businesses

Local shop owners are using AI to run targeted marketing campaigns. Imagine a bakery using AI to design promotional emails, suggest trending recipes, or predict which pastries will sell best during the holidays. AI gives them access to insights and strategies that were once only affordable for big corporations.

Writers and Creatives

Many independent authors, artists, and designers use AI to brainstorm ideas, create rough drafts, or generate images. Instead of stifling creativity, these tools often spark new directions and make it easier to bring ideas to life quickly.

Farmers

AI is transforming agriculture in remarkable ways. Farmers are now using AI-powered drones and sensors to monitor crop health, detect pests, and measure soil moisture in real time. Predictive AI models can forecast weather patterns, helping farmers decide the best time to plant or harvest crops. Some even use autonomous tractors and robotic pickers to save labour costs. AI can also analyse satellite data to suggest which parts of a paddock need more fertilizer or water, reducing waste and improving yields.

In short, AI is giving farmers precision tools that increase productivity, sustainability, and profit.

Builders and Tradies

AI is quietly reshaping the building and construction industry. Smart design software can generate 3D building plans, calculate materials, and even flag potential structural issues before work begins. On-site, AI-powered drones and scanners can monitor progress, measure accuracy, and check for safety hazards. Electricians, plumbers, and carpenters are starting to use AI tools that diagnose faults, estimate job costs, or suggest the most efficient layouts for wiring and piping. Even scheduling and supply ordering can be automated with AI, saving time and reducing costly errors.

Rather than replacing tradies, AI is becoming a handy tool on the job — the ultimate digital apprentice.

These examples show that AI isn't some distant or abstract force—it's here now, quietly reshaping everyday work and opening new possibilities for those who embrace it.

LOOKING FORWARD

The AI future isn't about machines replacing people—it's about people who know how to use machines replacing those who don't. By staying informed, experimenting with tools, and cultivating a flexible mindset, anyone can turn the AI revolution from a threat to an opportunity.

History reminds us that every major technological shift—from the steam engine to the internet—created winners and losers. The winners were not necessarily the smartest, but the most adaptable. The same will be true with AI. Those who embrace it, learn it, and use it, will thrive in the decades ahead.

KEY TAKEAWAYS – QUICK CHECKLIST

- *Learn the basics of AI tools: ChatGPT, image generators, AI search.*
- *Stay curious: Keep learning through articles, courses, and experiments.*
- *Build skills AI can't replace: Communication, creativity, empathy, leadership.*
- *Experiment with AI in everyday work and hobbies.*
- *Anticipate change in your field and prepare early.*

LIVING IN AN AI WORLD

Artificial Intelligence isn't just something that lives inside ChatGPT, self-driving cars, or science-fiction headlines. It's already woven through the everyday fabric of life — hidden in your phone, your streaming service, your car, even your supermarket trolley.

Most of the time, you don't *see* it working. You simply notice that things *"just happen"* a bit faster, a bit smarter, and sometimes a bit eerily on cue. Your photos organise themselves. Your playlists predict your mood. Your emails sort into tidy folders. The magic isn't magic at all — it's AI running quietly in the background.

We're living with an *invisible assistant,* one that doesn't have a face or voice most of the time but still influences how we move, buy, watch, read, and connect.

AI THE INVISIBLE ASSISTANT
Here's how AI can act as your invisible assistant:

SMART HOMES, SMARTER LIVING

The Connected Home
Adam and Mia's smart home adjusts lighting, temperature, and security automatically. When the forecast shows a heatwave, blinds close and the air-conditioning shifts to energy-saving mode.

How Smart-Home AI Learns
Machine learning observes daily habits and predicts preferences. Its comfort powered by data — but this raises privacy questions about who owns that data.

HEALTH IN YOUR POCKET

The Health Watch That Saves Lives
Barry's smartwatch detected a dangerous heart rhythm at 2 a.m., sending him to hospital before serious trouble struck.

How Wearable AI Works
Algorithms trained on millions of heartbeats, identify anomalies and alert users instantly. It's personal healthcare in real time — but it's only as safe as the privacy protecting it.

GETTING THERE FASTER

The Commute That Plans Itself
Sofia's navigation app predicts traffic and reroutes her automatically. AI analyses millions of sensor readings in real time, so she arrives on time.

From GPS to Predictive Mobility
Predictive analytics now power traffic systems, logistics, and airline scheduling. The next step: *cars that communicate with each other.*

SHOPPING, ENTERTAINMENT AND EVERYDAY CHOICES

Example: The Streaming Shortcut – Netflix knows Ethan's taste in 1990s crime dramas from thousands of viewing signals.

Example: The Grocery App – Lisa's supermarket app knows she is a vegetarian and remembers her weekly orders and even suggests specials.

Tech Insight: Why Recommendations Feel Personal – Collaborative filtering compares user preferences at scale, matching your habits with millions of others. It saves time but risks *"filter bubbles"* that limit discovery.

AI AT WORK WITHOUT YOU KNOWING

Banks, supermarkets, and utilities all rely on AI for fraud detection, inventory control, and spam filtering. Most of us interact with AI dozens of times each day — often without realising it.

EDUCATION, FITNESS, AND FINANCES

The Personal Tutor
Chloe's AI study assistant adapts lessons to her needs.

The Fitness Partner
Sam's running app monitors his stride and gives coaching feedback.

The Finance Guardian
Budgeting apps like *Frollo*, flag suspicious charges and manage spending in real time.

When Convenience Becomes Dependence
The invisible assistant saves time but can make us dependent. When AI removes friction, it can also remove reflection. We must stay aware of when we're choosing convenience over control.

THE PRIVACY PUZZLE

Example: The Smart Speaker That Listened Too Much
A Perth family recently found their device had recorded a private chat. Even small glitches can expose how thin the line is between help and intrusion.

Stronger data laws are emerging, but awareness remains our best defence.

WHEN AI GETS IT WRONG
Navigation apps steer cars onto closed roads; facial recognition misidentifies innocent people. Each mistake reminds us that AI models the world but doesn't truly understand it. Human oversight remains essential.

BALANCING BENEFITS AND BOUNDARIES

Ask yourself these questions:

- *Is this tool saving me time or stealing my attention?*
- *What data am I giving away?*
- *Am I still the one making decisions?*

Used wisely, AI enhances life. Used blindly, it narrows it.

SUMMARY
AI has become the plumbing of modern life — invisible but essential. The people who thrive will use it consciously, not compulsively.

Future Watch: Ambient AI Everywhere
Ambient intelligence will soon surround us. Homes will anticipate comfort, cars will self-coordinate, and health sensors will detect illness before symptoms appear. Technology will feel less like something we *use* and more like something we *live within.*

What this means for you
Stay visible and informed. The more invisible AI becomes, the more important it is to understand how it works — and when to switch it off.

GETTING MORE OUT OF CHATGPT

Chances are you're already using AI tools like ChatGPT on a regular basis — for writing, planning, learning, or simply satisfying your curiosity. But even seasoned users often miss out on the *best* results, simply because of how they phrase their questions.

In this next section, we'll explore how to get more useful, accurate, and creative responses from AIs, by improving the way you ask for them. With a few simple techniques, you can turn everyday prompts into powerful tools for thinking, creating, and solving problems.

Please note: *the following examples focus on ChatGPT; however, the same principles and techniques apply to most other AI systems.*

HOW TO WRITE PROMPTS LIKE A PRO
If you've ever typed a question into ChatGPT and thought, *"That's not what I meant at all,"* you're not alone.

Many users spend more time rewording their questions than actually using the AI's answers. The difference between a vague prompt and a great one, is like the difference between chatting with a stranger and briefing a skilled assistant who knows your goals.

When you learn how to write prompts like a pro, ChatGPT stops being a novelty and starts becoming a serious productivity partner. Whether you're a business owner, a teacher, a student or simply someone who wants to save time, the key is learning how to give AI the right direction — clearly, concisely, and with purpose.

Let's walk through how to do that.

WHY PROMPTING MATTERS

AI tools don't *"think"* like humans — they interpret instructions. If your request is too broad, you'll get general, wordy answers. If your request is too short, you'll get something that feels unfinished or off-topic.

Strong prompts solve this problem. They:

- *Give **context** (so the AI knows your situation)*
- *Define **intent** (so it knows what you want)*
- *Set **structure** (so the answer is formatted how you need it)*
- *Include **tone and audience** (so it sounds right for your purpose)*

Once you understand these building blocks, you can get clear, useful results almost every time.

Think of ChatGPT as your assistant — *not a search engine.*

- *When you use Google, you type keywords.*

- *When you use ChatGPT, you give instructions.*

That small mental shift changes everything.

Instead of typing *"marketing ideas for small business,"* try this:

"Act as a small business marketing consultant. Suggest five creative, low-cost marketing ideas for a local café wanting to attract new weekend customers".

See the difference?

The second version gives context, defines the role, and sets an expectation for the output. That's prompting like a pro.

THE 5-STEP FORMULA FOR GREAT PROMPTS
Let's break down a repeatable process that works for almost any situation.

1. Start with Context
Tell ChatGPT who you are and what you're trying to do.

Example:

"I'm a financial coach creating a one-page guide for retirees about managing superannuation".

This sets the scene and instantly gives the AI a purpose.

2. Define the Role
Give ChatGPT a *"hat"* to wear — it helps shape tone and expertise.

Example:

"Act as an experienced financial educator who specialises in retirement planning".

This ensures the style matches your audience. You can use roles like *copywriter, teacher, coach, data analyst,* or *event planner* — whatever fits.

3. Be Specific About the Task
Vague instructions get vague answers.

Instead of saying:

"Write something about staying healthy,"

try:

"Write three short motivational quotes about staying healthy after 60, suitable for a Facebook post".

Now the AI knows **what** you want, **how many**, and **where** it will be used.

4. Set the Format
If you want a table, list, summary, or script — say so.

Example:

"Summarize the following text into bullet points, each under 15 words".

Formatting instructions save huge amounts of editing time.

5. Review and Refine
Good prompting is a conversation, not a one-shot command. After reading the first output, ask:

"Can you make it sound more conversational?"

or

"Can you expand on point three with a practical example?"

Each refinement teaches ChatGPT more about what you want. Within a few exchanges, the results become sharper and more tailored.

FROM AVERAGE TO EXCELLENT

Basic Prompt:

"Write an email to promote my new online course".

Improved Prompt:
"Act as a professional email marketer. Write a short, persuasive email introducing my new online course on time management for freelancers. Focus on saving time, reducing stress, and increasing income. End with a clear call to action. Present it in email format with subject line and closing".

Result:
The first prompt gets you a bland, generic message.
The second gives you a ready-to-send marketing email that sounds professional and fits your audience.

BONUS: ASK FOR EXPLANATIONS
If you want to *learn* how to prompt better, ask ChatGPT to explain itself. After it produces something you like, type:

"Explain why you structured it this way and how I could improve my prompt next time".

This turns every session into a mini lesson on communication and clarity.

COMMON PROMPTING MISTAKES

Being too vague:
"Write something about AI".
(AI what? For whom?)

Overloading the prompt:
300 words of mixed instructions. Keep it short and focused.

Skipping the role:
"Act as a journalist," "Explain like a teacher," or *"Think like a designer"* all steer tone and vocabulary.

Forgetting the format:
Always specify list, paragraph, script, etc.

Not iterating:
You can almost always improve results with one or two follow-up prompts.

THE SECRET: CLARITY + CURIOSITY
Think of prompt writing as a mix of briefing and brainstorming. You're briefing the AI like you would brief a colleague — but you're also exploring possibilities.

You might start with:
"Suggest 10 creative book titles for a guide to buying a new car".

Then follow with:
"Make them shorter and funnier,"

or

"Now rework them to sound more professional".

The best results often come after two or three refinements — not the first attempt.

PROMPTING LIKE A PRO IN EVERYDAY LIFE
Once you get comfortable with it, this approach applies to anything. *e.g.:*

- *Planning a trip*
- *Writing an email*
- *Summarising a meeting*
- *Creating a social post*
- *Brainstorming headlines*
- *Drafting a speech*

Each time, remember:

Context → Role → Task → Format → Refine.

KEY TAKEAWAYS

- *Give ChatGPT context before asking a question.*
- *Assign it a role or perspective.*
- *Be specific about what you want and how you want it presented.*
- *Review the result, then iterate to improve it.*
- *Treat prompting as a conversation, not a command.*

Do that, and you'll quickly move from *guessing what to type* to *getting exactly what you need.*

Once you learn to prompt like a pro, ChatGPT becomes more than a tool — it becomes an extension of your own thinking.

ADVANCED PROMPTING TECHNIQUES

Once you've mastered the basics of writing clear, focused prompts, it's time to step up a level. Advanced prompting isn't about making your instructions longer — it's about making them *smarter*. Think of it like moving from giving directions to a taxi driver, to briefing a project manager who understands your long-term goals.

These techniques will help you unlock ChatGPT's full potential — whether you're brainstorming, writing, analysing, or planning.

MULTI-STEP PROMPTING:

Breaking Big Tasks into Small Wins
Instead of trying to get the perfect answer in one shot, professional users guide ChatGPT through a sequence of steps.

Example:

>   **Step 1:** *Summarise this article in three key insights.*
>   **Step 2:** *Turn those into a 100-word LinkedIn post.*
>   **Step 3:** *Suggest a catchy headline.*

Each step builds on the last. The AI stays focused, and you get better-structured, more refined results.

Tip: Treat ChatGPT like a collaborator — not a vending machine. Each step adds clarity.

CHAIN-OF-THOUGHT PROMPTING:
Asking the AI to *"Think Out Loud"*

When you want deeper reasoning — like problem-solving or planning — ask ChatGPT to show its thought process.

Example:

"Explain your reasoning step-by-step before giving the final answer".

This approach helps the AI check its own logic and often produces more accurate or creative outcomes.

Use it for:

- *Strategic decisions — Outline the pros and cons before recommending the best option.*
- *Research or analysis*
- *Troubleshooting complex issues*

CONTRAST PROMPTING:

Exploring Options Before Choosing One
Instead of asking for a single answer, ask for *alternatives* — it's a powerful way to explore tone, style, or structure.

Example:

"Write three short product descriptions: one formal, one casual, and one humorous".

Now you can compare and blend the best elements from each version.

Use it for:

- *Marketing copy*
- *Social media posts*
- *Speeches or headlines*
- *Naming or branding ideas*

Tip: Follow up with *"Now combine the best elements of version 2 and 3"* — this creates an even stronger final draft.

PERSONA STACKING:
Combining Multiple Roles
You've already seen how role-based prompts *("Act as a teacher," "Act as a journalist")* improve tone and accuracy. Advanced users take it a step further by combining roles — this is called **persona stacking.**

Example:

"Act as a marketing strategist and behavioural psychologist. Create a landing-page headline that appeals to both logic and emotion".

The first role adds structure, the second adds persuasion. You can mix roles like *editor + historian, coach + storyteller,* or *engineer + designer.*

The results often feel more nuanced and human.

PROMPT CHAINING:

Linking Conversations Together
When working on long projects (like a book, course, or campaign), save your progress and build from earlier prompts. Each stage becomes a foundation for the next.

Example workflow:

- *Brainstorm 10 chapter ideas.*
- *Pick the best three.*
- *Ask ChatGPT to outline each.*
- *Turn one outline into a 1,000-word draft.*

You're guiding the AI through a pipeline of refinement — just like a human editor would.

REFERENCE PROMPTING:

Teaching ChatGPT What *"Good"* Looks Like
You can feed examples into a prompt, so the AI understands your preferred tone, format, or structure.

Example:

- *Here's a sample of how I like my articles written: paste in example article.*
- *Write a new version of this story in the same tone and style.*

This works beautifully for:

- *Email templates*
- *Blog articles*
- *Reports or proposals*
- *Book chapters*

It's like training a personal writing assistant without needing coding or setup.

PRECISION PROMPTING:

Using Constraints for Sharper Results
Adding limits often produces better quality. You can specify word counts, tone, audience, or even forbidden phrases.

Examples:

- *Summarise in 75 words.*
- *Explain this for a 12-year-old audience.*
- *Avoid jargon or buzzwords.*

Constraints sharpen focus — for both the AI and the reader.

ROLE REVERSAL:

Ask ChatGPT to Interview *You!*
A fun but powerful technique: instead of giving instructions, invite the AI to guide you.

Example:

"Act as a business coach. Ask me five questions to clarify my goals before suggesting a marketing strategy".

This turns the conversation interactive. The AI becomes a smart interviewer — ideal for brainstorming, self-reflection, or refining an idea you're still shaping.

ITERATIVE REFINEMENT:

Layering Prompts for Perfection
Professionals rarely accept the first output. They treat the AI as a co-editor — refining through targeted follow-ups.

Example:

"Good start — now make the language more persuasive".
"Add a story example".
"Make it sound more realistic".

Each iteration adds polish and personality.

Tip: Save your best sequences. They become reusable *"prompt templates"* for future projects.

PROMPT LIBRARIES AND TEMPLATES
Once you find prompts that consistently produce great results, keep them. Organise them by category — *writing, marketing, planning, research,* etc. Over time, you'll build a personal toolkit that saves hours of effort.

Example template:
Act as [role]. **I am** [context]. **Your task is to** [goal]. **The output should be** [format]. **Tone:** [style].

Plug in the blanks each time. It's like having a Swiss Army knife for AI interactions.

COMMON MISTAKES TO AVOID

- *Asking too many things in one prompt*
- *Forgetting to specify tone or audience*
- *Not reviewing the reasoning (especially with factual tasks)*
- *Ignoring the chance to iterate or request alternatives*

Prompting is a skill — not magic. But with practice, it becomes second nature.

KEY TAKEAWAYS

- *Break complex tasks into multiple steps.*
- *Ask ChatGPT to explain its thinking.*
- *Compare options before choosing the best.*
- *Combine roles for richer output.*
- *Save and reuse your best prompt templates.*

Mastering advanced prompting turns ChatGPT from a clever assistant into a creative partner — one that learns, adapts, and grows with you. You'll find that the more intentional you become with your prompts, the more intelligent the AI seems in return.

Or as one pro user put it:

"The better I get at prompting, the smarter ChatGPT appears to become — but really, it's me who's getting smarter!"

TEACHING CHATGPT YOUR STYLE

Once you've learned how to prompt like a pro, the next step is making ChatGPT *sound like you.* Think of it as training a new team member — the better you explain your style, the faster it starts producing work that feels natural and authentic.

ChatGPT doesn't automatically know your tone, preferences, or quirks. But with a few clever techniques, you can teach it to write, think, and communicate in a way that mirrors your own voice — whether that's friendly, formal or professional.

WHY STYLE MATTERS
When you read something written in your own style, it *feels right.* Your rhythm, your vocabulary, your attitude — that's what connects you to your readers.

If you just say, *"Write an article about AI,"* you'll get something that sounds generic. But if you add a few lines about your personality, audience, or brand, the result changes.

Example:

"I write for people who are curious about AI but don't want the jargon. My tone is conversational, practical, and encouraging — a bit like talking to a smart friend over coffee".

Give ChatGPT that kind of direction, and it will instantly adjust its language and structure to match.

START BY SHOWING, NOT TELLING
The fastest way to teach ChatGPT your style is to give it examples.

Example Prompt:

"Here's a sample of my writing style":
(Paste a few paragraphs of your own work).

Describe the tone, vocabulary, and structure in detail so you understand it. This step helps ChatGPT *"analyse"* your writing.

It might respond with something like:

"Your tone is conversational, confident, and informative. You use short paragraphs, active verbs, and occasional humour".

Perfect. Now it has a clear reference point.

LOCK IN THE VOICE
Once ChatGPT has analysed your example, tell it to *adopt* that style.

Example:

"Use that same tone and structure for everything we write from now on. Confirm that you understand my writing style before we continue".

This simple command effectively sets your style as a baseline for all future responses in that conversation.

Tip: If you start a new chat, just paste your style summary at the top again — it's like reminding your assistant how you like things done.

USE A STYLE SHEET OR PERSONA SUMMARY
Professional writers and marketers often use a *style sheet* — a quick reference that defines their tone, word choice, and personality traits.

You can make one for yourself in ChatGPT.

Example:

MY VOICE GUIDE:

> **Tone:** *Friendly, knowledgeable, and conversational*
> **Audience:** *Curious readers aged 35–65*
> **Style:** *Clear, practical, slightly witty*
> **Avoid:** *Technical jargon and overused buzzwords*
> **Favourites**: *Short paragraphs, plain English, relatable examples*

Once you've written it, paste it at the top of any session and say:

"Use this voice guide for all responses, unless I say otherwise".

That's all it takes to make your writing instantly consistent.

TRAIN CHATGPT TO BE CONSISTENT
If the AI's output doesn't feel right, guide it with contrasts.

Example:

"This version sounds too formal. Make it warmer and more natural".

or

"That's too casual — make it sound more professional but still friendly".

These gentle corrections work like feedback during staff training. After two or three rounds, ChatGPT will lock onto your preferences with surprising accuracy.

USE REFERENCE MATERIAL
If you have a blog, newsletter, or book, you can feed in snippets from them to reinforce your tone.

Example:

"Here are three short excerpts from my writing. Use these as tone references when writing future chapters".

The AI will start picking up your rhythm, pacing, and choice of words. It's a quick way to build a consistent author voice across projects.

SET RULES FOR CONSISTENCY
If you write for a brand or publication, you might want specific guidelines.

You can give ChatGPT clear *"rules"* to follow.

Example:

"Always use English spelling and idioms".
"Avoid American slang".
"Use short sentences and no emojis".

By making expectations explicit, you save time editing later — and get writing that feels like *you* from the start.

TEACH CHATGPT TO WRITE FOR DIFFERENT AUDIENCES
You might have more than one tone — say, one for professional articles and another for newsletters or social media.

You can define each one and switch between them easily.

Example:

"When I say, 'write in my formal style,' use this tone: professional, balanced, and objective".

"When I say, 'my casual style,' write as if speaking to a group of friends over coffee".

Now you can say,

"Write this email in my casual style,"

and it will automatically adjust.

Combine Style with Role Prompts
You can merge your voice with a role instruction for more targeted results.

Example:

"Act as a professional editor but write in my conversational style. Make this paragraph sound clear, confident, and warm — not corporate".

This gives you the best of both worlds: professional structure with your personal flair.

Keep Iterating
Even with a solid voice guide, ChatGPT improves the more you interact. Don't hesitate to correct it when it drifts.

Example:

"You've made this a bit too formal — can you make it sound more like me again?"

or

"Add a touch of humour but keep it natural".

Each correction teaches the model more about your preferences. Over time, it becomes almost second nature for it to write *"like you".*

BONUS: TEACHING CHATGPT VISUAL STYLE
If you use ChatGPT to create images (like social media posts or marketing graphics), you can also describe your *visual* taste.

Example:

"My preferred design style is clean, minimal, and modern — with white space, soft shadows, and blue accents".

You can reuse that prompt whenever creating visuals, ensuring a consistent look across projects.

COMMON MISTAKES TO AVOID

- *Not giving examples — ChatGPT can't guess your tone.*
- *Mixing styles in one chat — it confuses the model.*
- *Forgetting to restate your voice guide in new sessions.*
- *Asking for opposite things — e.g., "formal but funny".*

Be clear, stay consistent, and your results will improve dramatically.

KEY TAKEAWAYS

- *Feed ChatGPT examples of your writing style.*
- *Ask it to analyse, summarise, and adopt your tone.*
- *Use a short "voice guide" at the start of every session.*
- *Give clear feedback when the tone isn't quite right.*
- *Save time by keeping different style templates for different audiences.*

Once ChatGPT learns your style, it becomes more than a writing assistant — it becomes your creative partner.
It can draft articles, emails, or chapters that sound like *you* on your best day: clear, confident, and engaging.

USING AI AT WORK —
PROMPTS FOR PRODUCTIVITY

Artificial intelligence is quietly becoming the world's most reliable office assistant. Whether you work in an office, run your own business, or manage a team, AI tools can draft emails, summarize meetings, organize your week, and even help you make smarter decisions.

Below are some real examples of how you can use ChatGPT for work — adapted from the *OpenAI Academy.* *
Each prompt includes an example of how ChatGPT might respond, so you can see it in action.

* The OpenAI Academy is a free educational service from ChatGPT to help people use AI. See chapter called: Upskill Your Knowledge of AI

REWRITE FOR CLARITY

Prompt:
Rewrite the following text so it is easier to understand. The text will be used in a professional setting. Ensure the tone is clear, respectful, and concise.

Text: *"As per our recent conversation, we must remind staff to ensure that all documentation is correctly formatted and uploaded before end of day Friday, as per the compliance requirement".*

Example ChatGPT Response:

Please make sure all documents are formatted correctly and uploaded by Friday's end of day to meet compliance requirements. Thank you for your attention to this.

WRITE A PROFESSIONAL EMAIL

Prompt:

Write a professional email to a supplier. The email is about delayed delivery of products and should be polite, clear, and concise. Provide a subject line and a short closing.

Example ChatGPT Response:

Subject: Update on Delivery Schedule for Recent Order

Dear Jamie,

I hope you're well. I wanted to check on the status of our recent order (#4721) placed on October 1st. We were expecting delivery by the 10th but haven't received confirmation of shipment yet.

Could you please advise when we can expect the products to arrive? We appreciate your help in keeping us informed so we can plan accordingly.

Best regards,

Your Name
Operations Manager

ADAPT A MESSAGE FOR AN AUDIENCE

Prompt:

Reframe this message for executives. The message was originally written for team members.

Text: *"We've completed testing on the new platform and fixed several bugs. We'll launch to customers next week".*

Example ChatGPT Response:

The product team has successfully completed final testing of the new platform and resolved key technical issues. We're on track for a public launch next week, as scheduled.

MEETINGS & COLLABORATION

Draft meeting invite

Prompt:

Draft a meeting invitation for a session about quarterly sales performance. The meeting will include regional managers and should outline agenda items, goals, and preparation required. Provide the text in calendar-invite format.

Example ChatGPT Response:

Subject: Quarterly Sales Review —
Regional Managers Meeting

When: Tuesday, 10:00 – 11:30 a.m.

Where: Zoom link to follow

Agenda:

- *Review Q3 results vs. targets*
- *Discuss regional challenges and highlights*
- *Set Q4 priorities and initiatives and note key learnings for discussion.*

You don't need to be a tech expert to benefit from AI at work. These examples show that with clear prompts and context, ChatGPT can become your personal assistant — helping you communicate better, organize more efficiently, and make smarter decisions.

The secret is not just using AI but teaching it what you need. Every time you clarify your goal, audience, and tone, you make the AI more useful — and your workday a lot easier.

CREATE A MEETING AGENDA

Prompt:

Create a structured agenda for a meeting about improving customer support response times. The meeting will last 60 minutes and include support managers and product leads.

Example ChatGPT Response:

Meeting Agenda:

Customer Support Response Improvement (60 min)

1. Welcome & Objective (5 min) — Clarify the goal: reduce average response time by 20%.
2. Current Metrics Review (10 min)
3. Root Cause Discussion (20 min)
4. Proposed Solutions (15 min)
5. Next Steps (10 min)

SUMMARIZE MEETING NOTES

Prompt:

Summarize these meeting notes into a structured recap. Organize them into categories: key decisions, next steps, and responsibilities.

Example ChatGPT Response:

Meeting Recap — Product Launch Planning

Key Decisions: Launch date set for May 1; marketing to focus on social media first.

Next Steps: Finalize pricing by April 10; confirm beta feedback session.

Responsibilities: Alice — marketing campaign; Ben — demo video; Carla — customer FAQs.

CREATE AN ACTION ITEMS LIST

Prompt:

Turn the following meeting notes into a clean task list grouped by owner.

Example ChatGPT Response:

Action Items:

John: Draft new client proposal (due Friday)
Emily: Update slide deck with feedback
Team: Review proposal and comment by Tuesday

ORGANIZATION & PRODUCTIVITY

DOCUMENT DAILY PRIORITIES

Prompt:

Create a prioritized to-do list from the following tasks: respond to client emails, finish report, attend meeting, update website.

Example ChatGPT Response:

Today's Priorities:

- *Finish report — Deadline approaching, high impact.*
- *Attend meeting — Key information.*
- *Respond to client emails — Maintain relationships.*
- *Update website — Lower urgency.*

CREATE A WEEKLY PLAN

Prompt:

Build a weekly work plan for a marketing manager preparing a new campaign.

Example ChatGPT Response:

> **Monday:** Research competitors, finalise messaging.
> **Tuesday:** Draft creative briefs.
> **Wednesday:** Meet design team.
> **Thursday:** Review campaign assets.
> **Friday:** Schedule posts, report summary.

WRITE A PROJECT UPDATE

Prompt:

Draft a short project update for stakeholders. The project is developing a new company website.

Example ChatGPT Response:

Subject: Project Update — New Website Development

The design phase is complete, and development is 60% finished. Testing navigation this week; launch on schedule for early December.

SUMMARIZE A LONG DOCUMENT

Prompt:

Summarize this 12-page report into 5 key points and 3 recommended actions.

Example ChatGPT Response:

Key Points:

- *Customer satisfaction +12%.*
- *Website engagement +30%.*
- *Mobile sales lagging.*
- *Delivery costs rising.*
- *Product returns steady.*

Actions:

- *Optimise mobile checkout.*
- *Negotiate courier contracts.*
- *Expand loyalty rewards.*
- *Brainstorm solutions*

Prompt:

Brainstorm potential solutions to reduce staff turnover.

Example ChatGPT Response:

- *Flexible scheduling — morale boost, coordination issue.*
- *Mentoring — loyalty, time cost.*
- *Pay review — competitive, costly.*
- *Recognition — motivational, consistency required.*
- *Exit interviews — insights, reactive.*

SUMMARY

These practical examples demonstrate how AI can support every aspect of modern work — from communication and collaboration to planning, analysis, and productivity. By learning to give clear prompts and context, anyone can turn AI into a trusted partner for better decisions and smoother workflows.

AI TOOLS THAT MAKE WORK EASIER

Artificial Intelligence isn't just about chatbots or coding assistants anymore — it's quietly reshaping the tools we use every day. From writing and design to scheduling and research, a growing range of AI-powered apps can help you get more done with less effort.

In this chapter, we'll explore some of the most useful categories of AI tools available — with examples of how they can make work faster, smarter, and even more enjoyable.

WRITING AND COMMUNICATION TOOLS

AI can take the stress out of writing — whether you're crafting emails, reports, or presentations.

POPULAR TOOLS:

ChatGPT – Ideal for drafting, summarizing, and improving tone.

Example Use: You type:

"Summarize this 2,000-word article into a one-paragraph executive brief".

ChatGPT produces a concise, professional summary in seconds — perfect for time-poor managers or clients.

Grammarly – Checks grammar, clarity, and tone in real time.

QuillBot – Rewrites text for simplicity or style.

Jasper – Geared for marketing and content creation at scale.

DESIGN AND VISUAL CONTENT TOOLS

You no longer need to be a graphic designer to create visuals that impress.

POPULAR TOOLS:

Canva Magic Studio
Create presentations, posters, or social media graphics using text prompts.

DALL-E
Generates original images from descriptions (great for book covers, ads, or blog posts).

Example Use:

Prompt:

"Create a professional image of an office team celebrating success, realistic style, landscape orientation".

DALL-E produces several versions instantly — no stock photo licensing or long searches required.

Runway ML
Edits and enhances videos with AI, removing backgrounds or generating effects.

Adobe Firefly
Integrates generative AI into Photoshop and Illustrator.

ORGANIZATION AND PRODUCTIVITY TOOLS
AI can take a lot of the effort out of managing your time and priorities. A few standout tools are already reshaping how people plan their days, organize tasks, and turn rough ideas into action plans.

Motion (or Reclaim.ai)
These apps automatically schedule your day based on your deadlines, priorities, and available time. If something changes—like a meeting being added or a task taking longer than expected—your calendar adjusts automatically.

Example Prompt:

"Plan my week to finish the project by Friday".

Motion blocks out your calendar for focused work, meetings, and breaks, so you stay on track.

Todoist AI
Todoist now includes an AI assistant that helps you prioritize or delegate tasks. It can suggest what to do first, reword vague items into clear actions, and even assign smart due dates.

Example Prompt:

"Review my task list and suggest what I should complete today based on deadlines and importance".

Todoist AI rearranges your list and highlights the most important items.

Notion AI
Notion AI helps you turn unstructured text—like meeting notes or brainstorming ideas—into organized action plans.

Example Prompt:
"Summarize my meeting notes and turn them into an action list grouped by owner".

In seconds, Notion AI structures your notes into clear responsibilities, deadlines, and next steps.

MEETINGS AND COLLABORATION

Meetings are often where time goes to die — but AI can help prevent that.

POPULAR TOOLS:

Otter.ai
Records and transcribes meetings automatically, creating searchable notes.

Fireflies.ai
Captures online meetings, summarizes discussions, and highlights decisions.

Krisp.ai
Removes background noise from calls for clearer audio.

Zoom AI Companion
Summarizes meetings, drafts follow-up emails, and suggests next steps.

Example Use:

During a meeting, Fireflies listens in, produces a full transcript, then sends a summary with bullet-point action items to everyone.

RESEARCH AND ANALYSIS TOOLS
AI tools can cut hours of searching and reading into minutes of insight.

POPULAR TOOLS:

Perplexity.ai
An AI search assistant that provides cited sources and summaries.

Elicit
Helps researchers summarize scientific papers and extract key findings.

Consensus
Uses AI to find evidence-based answers from peer-reviewed research.

ChatGPT with browsing
Can look up current information and synthesize it for you.

Example Use:

Prompt:

Find the three most recent studies on solar battery efficiency and summarize their key results.

Within moments, Elicit or ChatGPT can surface summaries with links to the original sources.

DATA AND BUSINESS INTELLIGENCE
Data-driven decisions are easier when AI does the heavy lifting.

POPULAR TOOLS:

Tableau Pulse (AI Insights)
Turns raw data into natural-language summaries.

Power BI Copilot
Generates visual reports and answers questions about your data.

ChatGPT (with CSV upload)
Analyses spreadsheets and produces charts or insights.

MonkeyLearn
Performs sentiment or text analysis on customer feedback.

Example Use:

Prompt:

Analyse this CSV file of customer feedback and tell me the top three recurring complaints.

MonkeyLearn instantly highlights the patterns — something that might take hours manually.

PERSONAL ASSISTANTS AND AUTOMATION
AI assistants are evolving beyond reminders — they now handle entire workflows.

POPULAR TOOLS:

Zapier AI / Make.com
Connects apps to automate repetitive tasks.

Clara or Motion AI
Acts as a scheduling assistant, booking meetings automatically.

ChatGPT Actions & Agents
Executes real-world tasks such as drafting reports or sending messages.

SaneBox or Superhuman
Filters and prioritizes your email inbox with AI.

Example Use:

When a client signs a contract, automatically send a welcome email and set up their project folder.

THE HUMAN FACTOR

The best AI tools don't replace people — they amplify them. AI takes care of the repetitive, administrative, or data-heavy work so that you can focus on creativity, relationships, and strategy. AI tools are assistants, not autopilots. They're most effective when guided by human judgment and ethical use. Think of AI tools as co-workers that never get tired or bored.

Start small — use one tool to solve one pain point. Always check the accuracy and privacy of any tool you use. The future of work belongs to those who learn how to delegate to machines — *wisely!*

EDITOR'S NOTE:

Listed above are just a handful of the thousands of AI tools currently available. For an extensive list of AI tools and what they do, visit this site: **Futurepedia.io**

Futurepedia.io is one of the most popular and reliable directories of AI tools available today. It lists thousands of applications across every category — writing, images, video, productivity, customer service, research, coding, and more and the site is updated daily as new tools emerge.

Each listing includes a clear description of what the tool does, making it easy for beginners and experienced users alike to explore what's possible with AI. If you want a simple, one-stop place to browse the ever-growing world of AI tools, *Futurepedia* is an excellent starting point.

GETTING BETTER RESULTS WITH AI IMAGES

How to master the nuances, not just the prompts

By now, most people using AI regularly have experimented with image generation — whether to illustrate a concept, create marketing material, visualise a new idea or just simply for fun. You've probably discovered that sometimes the results are stunning — and at other times, they miss totally.

The difference isn't random. It comes down to understanding how each image model interprets your words, where their strengths lie, and how subtle adjustments in phrasing can turn a mediocre result into a masterpiece.

This chapter is about going from good to consistently excellent. We'll explore how different image AIs *"see"* — how to push them for stylistic control, and how to maintain a consistent visual identity across your work.

UNDERSTANDING THE MAIN IMAGE MODELS — AND THEIR BIASES

Each AI image generator has its own style bias — its default *"look and feel"*. Think of them as different artists who each bring their own flair, even when following the same brief.

Each system was trained on different data, so the same prompt will yield different results — sometimes wildly so. Once you understand how they interpret prompts, you can lean into each model's strengths rather than fighting against them.

DISTINCTIVE TRAITS OF AI IMAGE TOOLS

AI Tool	Distinctive Traits	When It Excels
DALL-E 3 (in ChatGPT Plus)	Natural-language comprehension is unmatched — understands long, complex prompts and logical relationships. Produces clean, balanced, realistic scenes.	When you want accuracy, context, and coherence. Great for illustrations, editorial art, and concept scenes.
Midjourney	Highly stylised, cinematic, and artistic by default. It "interprets" creatively, often adding drama or mood even if not requested.	Perfect for storytelling, fantasy, and emotive or atmospheric images.
Adobe Firefly	Polished, brand-safe, and trained on licensed content. Integrates seamlessly with Photoshop and Illustrator.	Best for marketing, print, and corporate work where copyright matters.
Stable Diffusion / Leonardo.ai / Playground.ai	Open-source, deeply customisable. You can train models to reflect a consistent style or character.	Ideal for professionals who want full control or to build a recognisable brand look.
Google Nano Banana (Gemini 2.5 Flash Image)	Fast, flexible, and excellent at editing and compositing. It can blend photos, preserve identities, and maintain lighting consistency.	Outstanding for iterative edits, realistic image fusion, and facial consistency.

THE SMALL THINGS THAT CHANGE EVERYTHING

Experienced users already know the basics: describe the subject, the setting, and the style. But the next level comes from recognising how syntax and structure influence work.

WORD ORDER MATTERS
Most models weight earlier words more heavily.

"A photo of a futuristic city skyline at sunrise, detailed, cinematic lighting".

will often differ subtly from:

"Detailed cinematic lighting photo of a futuristic city skyline at sunrise".

So, front-load your most important concepts — the subject and the mood — before secondary details.

Avoid contradictions
Like combining conflicting visual styles.

i.e. *"photo-realistic watercolour painting"* dilutes intent.

Choose one dominant visual style per image.

Control creativity with modifiers
Use cues like *"precise" "clean"* or *"realistic"* for disciplined results. Or invite imagination with *"dreamlike," "artistic,"* or *"stylised".*

Reduce over-specification
Overloading a prompt with micro-details can make images stiff or cluttered. Instead of:

"A man wearing blue jeans, red hat, sitting on left side, holding green cup".

Try:

"A casually dressed man enjoying coffee in soft morning light".

Let the AI handle composition while you steer tone and mood.

UNDERSTANDING THE MODEL'S DEFAULT AESTHETIC
Just as human artists have signature strokes, AI models lean naturally toward certain aesthetics.

Midjourney loves mood — expect cinematic lighting, deep shadows, and bold contrast.

DALL-E 3 favours realism and logical storytelling.

Firefly defaults to bright, polished imagery with clean edges.

Stable Diffusion behaves like a blank canvas — it reflects the trained checkpoint or *"style model"* you select.

Nano Banana excels at realism and identity preservation; it tends to render natural skin tones and balanced lighting, even when blending multiple elements.

Knowing this saves time. If you want gritty realism, DALL-E 3 or Nano Banana will hit the mark faster than Midjourney, which naturally leans toward *"epic"* and *"artistic"*.

CONTROLLING CONSISTENCY ACROSS PROJECTS
For professionals, the challenge isn't just a single good image — it's producing many that look like they belong together.

Here's how experts do it:

Prompt Anchoring
Repeat exact phrasing for recurring features.
Example: if you want all images in *"soft golden-hour lighting,"* keep that phrase identical every time. Even punctuation can influence consistency.

Reference Images

Upload a base image (logo, product, face) and say:

"Create a new scene using this image as a visual reference — keep lighting and mood consistent".

Seed Values

Platforms like Midjourney and Stable Diffusion use a seed number to reproduce the same random noise pattern — essentially giving you *"version control"* over randomness.

Custom Models or LoRAs

For advanced users, you can train a Low-Rank Adapter (LoRA) or mini-model using your own sample images to reproduce a specific brand or character identity. That's how some publishers maintain consistent cover art styles across entire book series.

ITERATION: REFINING TO PERFECTION

The best image creators treat AI like a collaborator, not a one-shot generator. They refine, adjust, and iterate through feedback loops.

Generate broad variations first — explore style, lighting, and composition.

Select favourites and re-prompt the model with precise tweaks.

Example: *"Same image but slightly wider frame and softer contrast".*

Use image editing tools — DALL-E's *"edit image"* or Nano Banana's *"in-image revision"* features — to correct details rather than regenerating from scratch.

Outpainting expands an image (e.g., turning a square into a banner).

Inpainting repairs or replaces problem areas without touching the rest.

Firefly, Nano Banana, and DALL-E 3 all handle iterative editing well — with Nano Banana particularly strong in maintaining subject integrity across revisions.

IMAGE-TO-IMAGE CONTROL AND FUSION

This is where the magic happens. Instead of starting from scratch, you can guide the AI visually:

Upload a **sketch** or **photo**, and ask:

"Render this as a photorealistic image with soft studio lighting".

Combine two images:

"Blend this product photo with this background — make lighting consistent".

Reinterpret your own work:

"Turn this flat illustration into a cinematic version".

Nano Banana's fusion capabilities stand out here — it's adept at mixing multiple inputs coherently, while Firefly's *Generative Fill* gives designers pixel-level precision for replacements.

PROMPT LAYERING —
THE PROFESSIONAL'S SECRET WEAPON

Rather than writing a single sentence, advanced prompt engineers build prompts like a stack of layers, each defining a separate visual dimension.

Example structure:

Base concept: *"A bustling futuristic city at night".*
Style layer: *"Cinematic, ultra-detailed, reflective surfaces".*
Mood layer: *"Energetic, hopeful, luminous atmosphere".*
Technical layer: *"16:9 aspect ratio, sharp focus, 4K resolution".*

Each layer adds clarity without confusion. When results go off track, tweak only one layer at a time. This method works brilliantly across DALL-E, Midjourney, and Nano Banana alike.

ETHICS, ORIGINALITY AND ATTRIBUTION
As quality rises, so do questions about ownership and authenticity.

KEY PRINCIPLES FOR PROFESSIONALS:

Respect likeness: Don't use real people's faces without consent.

Label transparently: Many platforms now require declaring AI-generated content.

Use licensed models: Firefly and DALL-E 3 are trained on safe, rights-cleared datasets; Nano Banana embeds Google's SynthID watermark for transparency.

Avoid brand mimicry: Recreating known products or logos can infringe IP.

Good ethics build trust — both legally and reputationally.

THE NANO BANANA EDGE — WHERE IT SHINES
Google's Nano Banana deserves special recognition. It's part of the **Gemini 2.5 Flash Image** model family and represents Google's latest push into multimodal creativity.

WHAT SETS IT APART:

Face and object consistency: Great for series work where the same person or product must appear across multiple contexts.

Hybrid realism: Combines photographic accuracy with artistic flexibility.

Speed: Generations are near-instant and editable directly within the Gemini interface.

Fusion control: Merges multiple inputs with natural coherence — perfect for marketing composites or conceptual art.

In short: Nano Banana is the control freak's dream — less random flair than Midjourney and more adaptive than DALL-E. It's an excellent tool for professionals seeking dependable visual continuity.

THE ROAD AHEAD —
FROM IMAGES TO MOVING WORLDS
We're on the verge of another giant leap forward:

Text-to-video tools like Sora, Pika, and Runway already extend these image models into motion.

3D asset generation is becoming mainstream for AR and VR.

Voice-to-image input — describing visuals verbally — is now in prototype stages.

Tomorrow's creator won't just *"make an image"*. They'll describe a scene, a feeling, or even a moment, and AI will fill in every sensory detail.

FINAL THOUGHTS

At this stage, image generation isn't about pushing buttons — it's about directing vision.

You're not competing with AI; you're collaborating with it. The true skill lies in knowing how each system interprets your cues — how DALL-E reads logic, how Midjourney paints emotion, how Firefly polishes professionalism, how Stable Diffusion gives you freedom, and how Nano Banana quietly balances it all with realism and control.

When you understand those differences, you stop being an amateur user and start becoming a creative director of an intelligent visual team.

In the next chapter we'll look at the next big advance — creating videos with AI.

USING AI TO CREATE VIDEO

In this chapter we'll look at how text-to-video (and voice to video) AIs are redefining creativity.

Just a year or two ago, turning an idea into a video meant storyboards, cameras, actors, lights, editing software, a lot of patience and huge expense. Today, you can type a few sentences — or in some cases, simply speak them — and AI will generate a fully animated scene, complete with characters, movement, sound, and cinematic lighting.

Text-to-video and voice-to-video systems, represent the next leap in AI creativity. They bridge the gap between static imagination and moving reality. And while they're still in their early stages, progress is astonishingly fast.

This chapter explores what these tools can do, the differences between leading systems, and how to get the best results when directing your first AI-generated videos.

THE EMERGENCE OF GENERATIVE VIDEO

AI image generation felt magical when it first appeared — but adding motion multiplies the complexity. A single image contains millions of pixels; a video contains 24 frames every second that must all make sense in sequence.

Until recently, that was computationally impossible.
But new models — like *OpenAI's Sora, Pika Labs, Runway ML, Google Veo,* and *Synthesia* — are changing that by combining advances in large-scale diffusion, motion prediction, and multimodal understanding (text, image, and audio combined).

These tools don't just draw frames. They simulate physics, depth, lighting, and camera movement, then weave it all together into a smooth, believable sequence.

Here's how the main players differ today:

AI VIDEO GENERATION PLATFORMS

Platform	Core Strengths	Best For
OpenAI Sora	Generates high-definition video directly from text or still images. Exceptional realism, motion accuracy, and scene coherence.	Cinematic scenes, storytelling, concept visualization, advertising.
Pika Labs	Web-based, fast, and user-friendly. Converts text, stills, or short clips into stylised, looping videos.	Social media, short clips, creative reels.
Runway ML (Gen-2)	Powerful editing suite: text-to-video, video-to-video, inpainting, background replacement, and motion tracking.	Professional creators, marketers, YouTubers, and educators.
Google Veo (Gemini)	Produces detailed, long-form video from both text and voice input. Focuses on cinematic realism and colour accuracy.	Filmmakers and storytellers seeking continuity and tone control.
Synthesia / HeyGen	Specialise in avatar videos — talking heads delivering scripts in multiple languages.	Corporate training, online learning, explainer videos.

FROM PROMPT TO SCENE
The key to quality isn't magic — it's *direction*. AI video generation works best when your prompt sounds like a film brief.

Structure your prompt like this:

Scene description

- *Characters or objects*
- *Action or movement*
- *Camera style or mood*
- *Duration or format*

Example:

"A slow-motion drone shot of a surfer riding a huge wave at sunset, warm golden light, cinematic realism, 10 seconds".

You can add:

Camera directions

- *tracking shot*
- *handheld*
- *close-up*

Mood or style

- *melancholic*
- *documentary style*
- *1980s film look*

Framing

- *landscape for YouTube*
- *vertical 9:16 for social media*

Voice-to-video takes this a step further: you can simply describe the same prompt aloud, and the AI transcribes, interprets, and generates accordingly.

IMPROVING RESULTS
Even the best models benefit from iteration. Here's how to refine your outputs:

Start short
5- to 10-second clips render faster and reveal what works.

Use reference images
Upload a key frame or style example; tools like Runway and Sora adapt motion to match.

Mind transitions
If you plan to merge clips, describe start and end frames clearly.

Guide the lighting and camera
Phrases like *"natural sunlight" "steady camera,"* or *"slow pan left"* add realism.

Combine tools
Some creators use Midjourney or Nano Banana to generate key frames, then feed them into Runway or Pika to animate.

VOICE-TO-VIDEO — TALKING THE SCENE INTO BEING
Voice input is the next evolution. Instead of typing, you'll soon narrate a scene in natural language, and AI will render it on the fly.

Early prototypes (like Google Veo with Gemini voice integration and emerging versions of Sora) already demonstrate:

- *Real-time transcription and understanding of tone.*
- *Motion synced to vocal rhythm*
  (e.g., faster speech = faster pacing).
- *Ambient sound and voice-over blending automatically.*

It's easy to imagine creators directing entire sequences conversationally:

"Let's start with a close-up of the city skyline… now zoom out as rain begins to fall".

INTEGRATING AI VIDEO INTO YOUR WORKFLOW
AI video won't replace editing suites — it enhances them. Typical professional workflow:

- **Generate base footage** *(text-to-video).*

- **Edit** *in Premiere, Final Cut, or DaVinci for timing and audio polish.*

- **Enhance** *with AI tools — background replacement, lip-sync, or motion smoothing.*

- **Add narration** *with voice synthesis (e.g., ElevenLabs, Play.ht).*

- **Finish with branding** *in Canva or CapCut.*

For educators, marketers, and authors, this is revolutionary. One person can now storyboard, produce, and publish polished video without a crew.

ETHICAL AND LEGAL GROUND RULES
As video realism improves, responsibility increases.

- **Disclosure:** *Many platforms (YouTube, TikTok, Meta) now require labelling AI-generated video.*
- **Consent:** *Don't generate likenesses of real people without permission.*
- **Copyright:** *Avoid using prompts referencing existing film franchises or celebrities.*
- **Authenticity:** *Add visible watermarks or subtle identifiers to prevent misuse.*

Google's Veo and OpenAI's Sora embed invisible watermarks automatically; expect this to become standard.

WHAT'S NEXT?
In the coming year, expect:

- **Longer clips:** *from seconds to minutes.*
- **Real-time generation:** *live storytelling during meetings or classes.*
- **3D and AR export:** *your text-described scenes become immersive environments.*
- **Full multimodality:** *type, speak, sketch, or hum — the AI merges them all into unified video.*

This is where creativity becomes orchestration. You describe the idea once; AI handles the cinematography.

FINAL THOUGHTS
Text-to-video is more than a technical breakthrough — it's a shift in who gets to tell stories.

Small businesses, teachers, authors, and even hobbyists can now produce content that once required studios and big budgets. Soon, directing with your voice will be as common as typing a paragraph today. The winners won't be those who master software menus, but those who can think visually and describe their ideas clearly. So, next time inspiration strikes, try this:

Close your eyes, picture the scene, and simply tell your AI what you see. You might be surprised how quickly your words come to life on screen!

PROMPT EXAMPLES: FROM TEXT TO VIDEO
The best way to learn is by experimenting. Here are a few sample prompts you can try — or adapt — in tools like *Runway, Pika Labs, Sora,* or *Google Veo.* Each shows how a few extra words can change pacing, lighting, and emotion.

EXAMPLE VIDEO PROMPTS

Goal	Prompt Example	What You'll Get
Cinematic storytelling	"A drone shot flying over a rainforest canopy at dawn, mist drifting between trees, golden light breaking through clouds, cinematic realism, 15 seconds".	Smooth aerial footage with atmospheric light and natural motion.
Mood and atmosphere	"A city street at night in the rain, neon reflections, pedestrians with umbrellas, slow-motion tracking shot, 8 seconds".	Noir-style ambience with color reflections and depth.
Action and movement	"A surfer riding a massive wave, camera following from water level, splashes on the lens, high-speed footage".	Dynamic motion and perspective; water and spray effects.
Explainer or educational clip	"A close-up animation of DNA strands twisting and connecting, glowing in blue light, clean scientific aesthetic".	Smooth looping animation suitable for presentations.
Talking avatar	"Professional woman delivering a 30-second script in an office setting, natural hand gestures and lighting, clear English voice".	Corporate-style presenter video using avatar tools like Synthesia or HeyGen.
Historical recreation	"Old film look of 1920s streetcars moving through San Francisco, black-and-white grain, handheld camera effect".	Authentic vintage feel using texture and film artifacts.

Pro tip:
When you find a prompt that works — *save it* — and make small, deliberate variations (lighting, camera, mood). You'll quickly build your own prompt library of reusable cinematic recipes.

THE NUANCES OF PROMPTING FOR VIDEO

Prompting for video is as much about direction as it is about description. When you're generating images, you describe what you want to see.

With video, you're also describing what you want to happen. That means words like *"slowly," "suddenly," "tracking," "handheld,"* or *"steady drone shot"* can transform the same scene into completely different experiences.

Consider this example:

Prompt 1:

"A young girl walking through a forest, sunlight flickering through the trees".

Produces a calm, linear shot — she moves steadily through soft light.

Prompt 2:

"A young girl running through a dark forest, camera shaking slightly, flashes of light through branches".

The mood flips from peaceful to urgent. The camera motion becomes handheld and tense.

Or take this subtle variation:

Prompt 3:

"A car driving along a coastal road at sunset, cinematic lighting".

The AI may choose a wide, aerial view.

Prompt 4:

"Camera mounted inside the car, dashboard view as the sun sets over the ocean".

Now the scene becomes intimate — you're the driver, not the spectator.

Even the order of words can matter.

Putting the **action first:** *"a camera pans over..."* signals what to animate, while starting with **the subject:** *"a mountain range at sunrise..."* tells the AI what to focus on visually before motion.

When prompting for video, imagine you're briefing a film crew — not painting a still image. The clearer your direction of movement, the better the AI interprets transitions, rhythm, and perspective.

FIVE QUICK TIPS FOR BETTER VIDEO PROMPTS

Lead with motion.
Start prompts with what's moving — *"camera pans across"* *"waves crash against rocks"* or *"a drone flies over"* — to help the AI focus on dynamic flow rather than static composition.

Set the pace.
Words like *"slowly"* *"in fast motion,"* *"time-lapse"* or *"real time"* define rhythm and can completely change the emotional feel of a clip.

Specify the point of view.
"Wide aerial shot" feels detached; *"handheld close-up"* feels immersive. Choosing the viewpoint is like choosing the audience's seat in the scene.

Anchor lighting and mood.
Add context for tone — *"misty morning" "neon-lit street" "golden hour"* or *"overcast sky"* — to maintain continuity across clips.

Write like a director, not a camera.
Focus on intent: *"make it feel suspenseful" "create a sense of wonder"* or *"show the anticipation before the race begins".*

The AI will often interpret emotion cues better than overly technical instructions.

SUMMARY
I hope you found the tips above useful — and that you're inspired to experiment with creating your own videos using AI. Go ahead — *lights, action, camera!*

In the next section, we'll turn our attention to **AI Agents** — a fast-growing phenomenon that's reshaping how we use AI in everyday life. These agents are appearing everywhere, quietly taking on tasks that once required direct human input.

As the AI future unfolds, agents will become an increasingly important part of how we work, communicate, and stay organised.

THE AGENTIC FUTURE:
HOW AI AGENTS ARE BEING USED

An AI agent is like a smart digital assistant that can do things for you automatically — not just answer questions.

Think of it as a version of ChatGPT (or any other AI) that can take action on your behalf — rather than just talk.

For example:

You might tell your AI agent:

"Schedule a meeting with Bob next week and send him the agenda".

The agent will then check your calendar, email Bob, and even create the meeting invite — all by itself.

In short:

- *ChatGPT — answers questions.*

- *An AI agent — takes actions.*

AI agents can:

- *Search the web for you.*
- *Write and send emails.*
- *Book flights or restaurant tables.*
- *Analyse spreadsheets or reports.*
- *Even run ongoing tasks like "track AI news every week".*

They can work autonomously (on their own) or with your approval every step of the way.

HOW ARE AI AGENTS BEING USED?

Unlike traditional programs that wait for your input *("open email," "play music," "search the web"),* an AI agent has initiative. It can plan tasks, interact with other systems, and continuously learn from feedback.

Think of an agent as your *digital coworker* — not just a smarter chatbot, but an autonomous assistant capable of coordinating across tools — email, calendars, databases, web services — to get things done.

Recent advances in large language models (like GPT-5) have made this leap possible. Agents can now reason, retrieve live data and collaborate with humans in real-time. The line between *"assistant"* and *"partner"* is beginning to blur.

A DAY IN A LIFE WITH AI AGENTS

It's 7:00 a.m. Monday. Your personal AI agent has already been awake for hours. Overnight, it scanned your inbox, summarised key messages, paid your utility bill, and rearranged tomorrow's meeting so it no longer clashes with your doctor's appointment. It's also checked the weather, adjusted your car's charging schedule, and reminded your gym that you'll be skipping Pilates this week.

As you pour your coffee, a gentle alert pops up:

"Good morning, Carlo. You have a client call at 9:30, and I've drafted talking points based on last week's notes. Also, your new travel itinerary is confirmed — seat 12A as usual".

Throughout the day, your agent doesn't wait for you to ask for help — it anticipates your needs, books appointments, follows up on tasks, and handles the routine chores that normally chip away at your focus. When a friend's AI agent messages yours to set up lunch, the two quietly negotiate a time and location that works for both of you.

This isn't a futuristic dream; it's the emerging *agentic era* of AI. One where technology doesn't just *respond* to commands — *it acts on your behalf.*

WHERE ARE THEY BEING USED TODAY?

AI agents are already quietly transforming many industries — not through flashy headlines, but through practical day-to-day applications as follows:

Productivity and Workflows

Tools such as *Reclaim.ai* or *Motion* act like digital project managers, automatically scheduling tasks and reshuffling calendars to meet deadlines. Imagine your personal planner that never sleeps, tracks dependencies, and keeps everyone on schedule.

Customer Service

Businesses are deploying AI agents that can resolve refunds, update accounts, or handle routine queries without a human ever stepping in. The human team steps in only when emotion, complexity, or empathy are required — the perfect *"human-in-the-loop"* model.

Coding and IT

Developers now use AI coding agents that debug software, suggest optimisations, and even write deployment scripts. Instead of replacing engineers, these agents handle the repetitive work, letting coders focus on design and innovation.

Personal Finance

AI financial agents can analyse your spending, rebalance investments, and even negotiate better deals from your utilities or insurance. Rather than reacting to what you do, they take the initiative — acting more like a financial concierge than a simple budgeting app.

This shift toward *active assistance* is what makes agents different from older AI tools. They are not just informative; they are *operational*.

Challenges Along the Way

The road to an agentic future is not without potholes. As these systems gain autonomy, we face new questions around trust, mistakes, and security.

Trust and Transparency:

How much control should we hand over to software that makes decisions for us? Transparency about how and why an agent acts will be crucial.

Mistakes and Misunderstandings:

Even the best agents will occasionally misread intent or act prematurely. This is why many systems still rely on a *human-in-the-loop* model — where the agent drafts or recommends actions, but a human approves them before execution.

Security and Privacy:

Giving agents access to calendars, finances, and communications opens the door to risk. Companies are racing to build stronger safeguards, encryption, and ethical oversight to keep data safe.

Agents must earn trust the same way humans do — through reliability, clarity, and consistent performance.

COMMON MYTHS ABOUT AI AGENTS

Like any new technology, AI agents are surrounded by hype, confusion, and even fear. So, let's clear up a few of the biggest myths.

Myth 1: AI Agents Will Instantly Replace Human Jobs

The reality is more nuanced. Agents excel at narrow, repetitive tasks but struggle with judgment, empathy, or creativity. Most current use cases show agents working alongside people, not replacing them outright. Just as spreadsheets didn't eliminate accountants but changed their role, agents are more likely to shift jobs than erase them.

Myth 2: AI Agents Can Think for Themselves

Despite the headlines, AI agents are not conscious, self-aware, or capable of independent thought. They follow instructions, learn patterns, and optimise for goals — but they don't *want* anything. They are sophisticated tools, not digital humans.

Myth 3: AI Agents Are Already Fully Autonomous

Science fiction imagines AI running entire companies on its own. In reality, today's agents are carefully supervised. They operate best in constrained environments (like managing a calendar or handling returns for a retailer). Giving them wide-open autonomy

Myth 4: Only Big Companies Will Use AI Agents

At first, agents may seem like a luxury for large corporations. But just as email, websites, and cloud storage trickled down to small businesses and individuals, AI agents will become increasingly accessible. Startups, students, and even retirees will soon use personal agents in daily life.

TAKEAWAY

AI agents are moving us from a world where technology is *passive* to one where it's *active*. They don't just sit and wait for input — they *anticipate, plan,* and *act*.

83

The winners in this new world won't necessarily be those with the most powerful AI, but those who learn how to work best with it. The skill of the future will be *delegation*: knowing what to trust your agent with — and when to step in yourself.

FUTURE WATCH: THE RISE OF PERSONAL AI AGENTS

Over the next few years, personal AI agents may become as common as smartphones. They could:

- *Manage your inbox like a digital secretary.*
- *Negotiate schedules directly with other people's agents.*
- *Coordinate travel, finances, or even aspects of your health.*

In time, your agent might even interact with others — booking events, sharing data, or managing your household ecosystem automatically. What will matter most is not the technology itself, but the trust and relationship you build with your agent. The art of the future won't just be *using* AI — it will be *delegating wisely.*

THE RISE OF AI COMPANIONS

Personal AIs are evolving from assistants to partners. It started with simple chatbots — the kind that gave you weather updates or answered trivia questions. Then came voice assistants like *Siri* and *Alexa*. Now, we're entering a new era: *the age of AI companions* — intelligent digital partners that not only complete tasks but understand you, adapt to you, and grow with you.

This shift marks one of the biggest leaps in human-technology relationships since the smartphone. It's not just about automation anymore — it's about connection.

FROM TOOLS TO TEAMMATES

Until recently, technology was mostly reactive. You pressed a button, and it performed an action. Even early AI systems like *Siri* or *Google Assistant* responded only when prompted. But as systems become *context-aware*, they begin to remember preferences, learn your communication style, and anticipate your next move.

An AI companion might say:

"You've seemed under pressure this week. Want me to shorten next Monday's schedule?"

That's not science fiction — it's already happening. These systems are learning to interpret tone, rhythm, and context, not just words. They're blending emotional intelligence with computational power.

In this new phase, your AI becomes less of a *"tool"* and more of a *"teammate"*. It knows your goals, keeps track of what matters to you, and helps manage the constant noise of modern life.

WHAT MAKES AN AI A COMPANION?

Three key traits separate an AI *companion* from a traditional *assistant*:

Memory and Context – It remembers prior interactions and uses them to provide continuity *("Last time you asked about investing, you were focused on ethical funds — would you like me to check the latest options?")*.

Personality and Tone – It adapts its style to suit yours. Serious, playful, formal, or casual — it mirrors your communication preferences naturally.

Proactive Empathy – It senses emotional cues and responds appropriately. If you sound tired, it might slow down the pace or suggest a break.

In other words, companions don't just do tasks — *they relate to you.*

REAL-WORLD EXAMPLES

Replika – One of the earliest emotional AI companions, designed to provide conversation, motivation, and even friendship. It learns from each chat and tailors responses to suit the user's personality.

Pi by Inflection AI – Marketed as a *"personal intelligence,"* Pi combines conversational warmth with factual assistance. It's less about work tasks and more about thoughtful conversation and coaching.

ChatGPT with Memory – The next generation of ChatGPT can remember details across sessions — your projects, writing style, and even your preferences — effectively becoming your personal collaborator.

Character.ai and **Digital Personas** – These platforms allow users to create characters (from historical figures to fictional friends) and interact in surprisingly humanlike ways, blurring the boundary between entertainment and companionship.

Each of these examples highlights a subtle but important change: *We're beginning to form relationships with AI.*

WHY PEOPLE ARE DRAWN TO AI COMPANIONS
Loneliness, remote work, and digital overload have made meaningful connection harder to find. For some, AI companions offer a sense of continuity and understanding that's missing in human-only interactions. But there's another reason — *efficiency.*

Companions can serve as mentors, motivators, or creative partners — people use them to practice languages, brainstorm ideas, or rehearse difficult conversations. They don't tire, judge, or lose patience.

As one researcher put it, *"AI companions don't replace human relationships — they fill the spaces between them"*.

OPPORTUNITIES AND CONCERNS

The rise of AI companions raises both exciting possibilities and serious questions.

OPPORTUNITIES:

Personal growth: digital mentors, health coaches, and emotional support companions.

Accessibility: voice-based companions helping those with disabilities or cognitive challenges.

Learning and creativity: a constant, patient partner to explore new ideas with.

Concerns:

Emotional dependence: There's a danger in people becoming overly attached to non-human entities.

Privacy: Companions that *"know you best"* also hold deeply personal data.

Manipulation and trust: If a company controls your companion, whose interests does it serve — *yours, or theirs?*

Balancing empathy with ethics will be the defining challenge of this new relationship.

THE NEXT STEP: INTEGRATED COMPANIONSHIP
Soon, AI companions won't live in a single app — they'll move *with you*. Imagine your companion syncing across your phone, car, watch, and home devices, always aware of what's happening and how to help. It might whisper directions through earbuds, brief you before a meeting, or quietly manage your digital world in the background.

These systems will increasingly use multimodal AI — combining text, voice, vision, and gesture — to make interactions more natural. Instead of typing, you'll simply talk, glance, or gesture — and your companion will respond like a friend who's been by your side for years.

TAKEAWAY
AI companions represent a new frontier — where technology meets emotion, and data meets empathy. They are not just smarter assistants but the early versions of *digital partners* that could one day understand us better than we understand ourselves.

The challenge — and opportunity — will be to shape these companions so they serve humanity's best interests. The goal isn't to build machines that *replace* human connection, but ones that enhance it.

In the near future, the question won't be *"Do you use AI?"* It will be: *"Who's your AI?"*

HOW TO SET UP AND MANAGE AI AGENTS

We've looked at what AI agents can do — and their limits.
Now let's get practical. If you want to experiment with your
own AI assistant or build one into your workflow, this chapter
walks you through the essentials.

Warning and Disclaimer
Before setting up or using AI agents, proceed with caution.
These systems often require access to your personal data,
accounts, or device settings to perform tasks — and that
access can carry risks. Always understand what permissions
you're granting, where your data is stored, and how it may be
used. Avoid connecting agents to sensitive information or
financial services unless you fully trust the source and
understand the safeguards in place.

START SMALL, LEARN FAST
Don't begin by handing an AI agent your entire digital life.
Start with a single, low-risk task — something repetitive and
measurable.

Examples:

- *Scheduling meetings or reminders.*
- *Summarising your inbox each morning.*
- *Drafting routine emails or social posts.*
- *Collecting data from spreadsheets.*

The idea is to observe how the agent behaves and how well it
interprets your intent before trusting it with anything sensitive.

CHOOSE THE RIGHT PLATFORM
A growing number of tools now let you build or use agents
easily, often without coding:

OpenAI GPTs
Customisable agents built inside ChatGPT, able to remember instructions and access specific files or APIs.

Reclaim.ai / Motion
Time-management agents that automatically schedule your day.

Zapier or Make (Integromat)
Automation hubs where you can link apps like Gmail, Slack, and Google Sheets to form simple *"if this, then that"* workflows.

Hugging Face Agents / LangChain / CrewAI
For the technically minded — frameworks for building more advanced autonomous agents with custom actions.

Tip: Start with consumer-friendly options first. Once you understand the behaviour and permissions model, you can move to more complex systems.

BE EXPLICIT WITH INSTRUCTIONS
Agents aren't mind-readers. They thrive on specific guidance. The more detail you give them, the fewer surprises you'll have.

Example prompt:

"Check my calendar for next week. Find any meeting longer than 60 minutes. Suggest two shorter time slots that would still include all attendees".

A vague prompt like: *"Fix my calendar"* could cause chaos. Clarity equals safety.

MANAGE PERMISSIONS CAREFULLY

Before connecting your email, files, or accounts, stop and review:

- *What exact data does this agent need?*
- *Where is the data stored?*
- *Can I revoke access easily?*
- *Does it share information with third parties?*

If an agent demands *broad* access *("full access to Google Drive")*, pause. See if you can narrow it down *("read-only access to one folder")*. Always test with dummy files before linking the real thing.

Golden rule: never connect an agent to financial accounts or private client data until you fully understand how it handles privacy.

USE HUMAN-IN-THE-LOOP SETTINGS

Look for systems that allow approval workflows — where the agent prepares a task but waits for your confirmation before execution.

Example:

An AI could draft an email, but you review it and send it. It might generate a report, but you approve it before distribution. This keeps you in control and ensures that mistakes are caught early.

LOG AND REVIEW ACTIVITY

Many modern agents offer logs showing what they've done and why. Check these regularly. It's the digital equivalent of reading your assistant's notebook.

If something looks odd — an action you didn't expect, or data pulled from the wrong source — investigate immediately. These checks build trust and help refine instructions.

KEEP SENSITIVE DATA OUT OF REACH
Even the best platforms can experience bugs or data leaks.

Avoid sharing:

- *Private identification details (passport, tax file number).*
- *Passwords or financial logins.*
- *Client records or legal documents.*
- *Anything you wouldn't email to a stranger.*

If your agent needs to work with such data, anonymise it or use encrypted systems designed for that purpose.

CREATE A *"KILL SWITCH"*
This is good practice: ensure you can pause or disable an agent instantly. Whether it's a rogue automation loop or just a mistake, you need the ability to shut it down quickly.

On most platforms, this means disabling its API key, turning off workflow triggers, or revoking permissions in your account settings.

KEEP HUMANS IN THE LOOP
The smartest users are those who *collaborate* with AI, not surrender to it.

- *Treat the agent as a junior colleague.*
- *Delegate repetitive work.*
- *Supervise and review outcomes.*
- *Provide feedback to help it improve.*
- *AI agents get better when guided by human oversight.*

The partnership — not the automation — is where the real productivity lies.

EVOLVE RESPONSIBLY

As you gain confidence, you'll find ways to chain agents together or integrate them into daily life — perhaps linking your personal assistant with a work scheduler or a travel planner.

Just remember with each new connection comes more complexity and risk. Scale thoughtfully. And always document what you've built so you can trace how data moves through your system

QUICK CHECKLIST FOR SAFE AGENT SETUP

- *Start with one simple task*
- *Choose a reputable, transparent platform*
- *Give the agent minimal permissions*
- *Be precise in your instructions*
- *Keep approval steps for key actions*
- *Check logs and outputs regularly*
- *Protect private data*
- *Know how to turn it off*
- *Create a "kill switch".*

SUMMARY

The agent revolution isn't about machines taking over — it's about *smart delegation*. Used properly, AI agents can save hours each week, cut errors, and give you back mental space for creativity and strategy. But safety, clarity, and control come first.

Start small, stay alert, and keep the human in charge. That's how you turn an AI agent from a novelty into a trusted partner.

AI ASSISTED BROWSERS

As discussed, AI agents can now take action on your behalf — scheduling meetings, sending emails, or tracking information automatically.

A closely related development is the rise of *AI-assisted browsers,* which bring that same intelligence into the way we explore the web. Instead of just showing you pages, these browsers can now *read, summarise, and respond* — turning everyday browsing into an interactive conversation.

For most of the internet's history, web browsers like *Chrome, Safari,* and *Firefox* have been passive tools. They opened pages, displayed results, and stored bookmarks — but they didn't really help you to *think, write,* or *decide.*

All that is now changing.

A new generation of *AI-assisted browsers* — such as Google Chrome with Gemini or Microsoft Edge with Copilot — combine the traditional web experience with the intelligence of AI systems. Instead of just loading web pages, they *interpret* them, *summarise* them, and *help you act* on them.

WHAT MAKES AN AI ASSISTED BROWSER DIFFERENT?
An AI browser combines *search, assistant,* and *automation* — all in one window.

For example, instead of typing *"best hybrid cars available"* and skimming dozens of results, you can simply ask:

"Compare the best hybrid cars under $50,000 and show me owner reliability data".

The AI browser will search multiple sources, summarise the findings, and often include citations *(as links to where it got the information from)* so you can verify them yourself.

Here's a chart of the major differences between a regular browser and an AI browser like *Chrome* or *Edge*.

Traditional Browser	AI-Assisted Browser
Displays web pages.	Understands what's on the page.
You search, click, and read.	You ask questions in plain English.
You copy/paste between tabs.	It can write, summarise, translate, and generate.
No awareness of your context.	Can learn your preferences and tasks over time.

WHAT THEY CAN DO
AI assisted browsers can do the following things:

- *Summarise long articles, research papers, or YouTube transcripts.*
- *Rewrite or simplify text directly on the page (handy for emails or social media posts).*
- *Generate content — such as posts, outlines, or reports — while referencing live web data.*
- *Cross-search multiple sites at once and provide unified answers.*
- *Explain complex concepts without leaving the page.*
- *Personalise results based on your ongoing interests or history.*

WHAT THEY CAN'T DO *(YET)*

- *They don't always access every source — many still rely on public pages, not paid or subscription databases.*

- *They can misinterpret context if a page is technical or ambiguous.*
- *They're still learning how to handle privacy and data storage responsibly.*
- *They aren't (yet) perfect replacements for dedicated research tools or human judgment.*

In short: AI browsers are *assistants — not authorities.*

They save time and effort, but their answers still need a quick *"sense check"* — just as you'd double-check a friend's summary.

THE BIG PICTURE

In the coming years, browsers may feel less like tools and more like companions. Instead of jumping between *Google, Wikipedia,* and *ChatGPT,* you'll simply interact with one intelligent workspace that reads, writes, and reasons alongside you.

Browsing won't be about finding information anymore — it will be about understanding it faster.

As AI becomes part of every browser, the skill will shift from *"searching well"* to *"asking well".* That means learning to phrase questions clearly, specify what you need, and review answers critically — the same prompting skills you've already been developing in this book.

KEY TAKEAWAYS

AI-assisted browsers mark the next step in everyday computing. They bridge the gap between information and action — between *searching* and *doing.*

Those who learn to use them thoughtfully will navigate the future web not just faster, but smarter.

TRY IT YOURSELF:
Here are some typical questions to ask your AI assisted browser:

Do a quick survey:
"Find three recent opinions from experts on the future of electric cars and summarise them".

Compare products or services:
"Compare the main features and prices of the top 3 AI-powered note-taking apps".

Check credibility:
"Who wrote this article and what is their background? Are they considered a reliable source?"

Analyse a trend:
"Summarise the latest news about how AI is changing healthcare — focus on key innovations and challenges".

Summarise a site:
"Give me a one-paragraph summary of this website — what does it do and who is it aimed at?"

Extract facts or data:
"List the main statistics mentioned in this report about renewable energy".

Write a quick summary for a presentation:
"Turn this web page into three bullet points I can use in a PowerPoint slide".

Find contrasting views:
"Show me arguments both for and against using facial recognition in public spaces".

Create a short reference list:
"Find five reputable sources on the economic impact of AI and summarise each in one line".

AGENTIC AI BROWSERS — A GLIMPSE INTO THE FUTURE

AI browsers are already changing the way we search and browse online but the next big leap may come from what are being called *agentic AI browsers.*

WHAT ARE AGENTIC AI BROWSERS?
A traditional browser lets you do the work — you type, click, read, and act. An AI browser goes a step further by helping you do those tasks faster, using built-in AI tools to summarise, explain, or even draft content for you.

An *agentic AI browser* takes this to yet another level. The word *agentic* means it can act as an *"agent"* on your behalf. Instead of simply helping you find or understand information, it can take action — such as filling out forms, booking flights, or conducting research over time — all without constant supervision.

One recent example is **Fellou**, *(fellou.ai)* an early agentic browser that claims to *"browse the web for you"*. Users might ask it to compare prices, analyse product reviews, or monitor news on a topic—and it will continue working in the background, reporting back when it finds results.

WHAT THEY CLAIM TO DO
Agentic AI browsers promise to make online activity effortless. They can:

- *Search and summarise multiple pages automatically*
- *Compare products or services and give recommendations*
- *Fill in web forms or make purchases*
- *Monitor a topic or task over days or weeks*
- *Learn your preferences and adapt their behaviour*

NON-AGENTIC AI BROWSERS VS AGENTIC

Feature	Non-Agentic AI Browser	Agentic AI Browser
Main Function	Enhances browsing with AI summaries and suggestions	Acts on your behalf—searches, clicks, fills forms, completes tasks
User Control	You stay in charge—AI assists but doesn't act alone	AI can take initiative with limited supervision
Example Tasks	Summarise articles, rewrite emails, explain content	Compare deals, book travel, track news, manage accounts
Risk Level	Low—AI works only within your instructions	Higher—AI may make errors or misjudge intent
Privacy & Security	Minimal data sharing	Requires access to accounts, logins, or payment details

THE LIMITATIONS AND DANGERS
While the idea sounds futuristic and appealing, it's important to be cautious. Agentic browsers are still in the experimental stage. For them to take meaningful action, they need deep access to your digital life — accounts, passwords, credit cards, calendars, and personal data. That opens the door to obvious risks if anything goes wrong.

Autonomous agents can also misinterpret instructions or act in unintended ways. A misplaced booking, an incorrect purchase, or the sharing of sensitive data could all happen without you realising it until after the fact.

Then there's the broader concern of *over-delegation*. The more we allow software to make decisions, the more detached we become from the choices themselves.

Convenience can quietly erode control. For most people, AI that assists is fine; AI that acts independently still demands caution.

EDITOR'S NOTE

The line between AI-assisted browsers and agentic browsers is becoming increasingly blurred. Many *"AI-assisted"* browsers are now adding agentic capabilities — for example, Chrome/Gemini currently states: *"agentic browsing is coming."*

At the same time, some agentic browsers still require a high level of user oversight. For instance, at the time of writing, OpenAI's Atlas browser requires you to manually opt in and grant permission for agentic actions. Other agentic browsers may not offer the same safeguards. If you decide to install an agentic browser, it's important to understand how much autonomy you're handing over.

A SENSIBLE APPROACH

For now, it's best to think of agentic browsers as an early preview of what's coming rather than something to rush into. The underlying technology — autonomous agents — will almost certainly shape future computing, but it's wise to stay informed without handing over the keys just yet.

Use AI browsers that assist and accelerate your work but keep a human firmly in charge of decisions involving money, security, or personal information.

The future may well belong to these digital agents, but for now, they're better treated as promising apprentices than trusted employees.

UPSKILL YOUR KNOWLEDGE OF AI

AI is moving at an amazing pace and to keep up with it, you need to adopt a policy of continuous learning. If you're serious about using AI in your work or education or you just want to stay ahead of the curve, there are now several ways you can do this, many at low cost or no cost.

JOIN THE OPENAI ACADEMY
the *OpenAI Academy* is one of the best places to start. It's free and open to everyone, regardless of whether you're using ChatGPT for free or as a paid user — no subscription required.

Once you sign up, you'll have access to structured content, tutorials, workshops, and community groups that help you deepen your AI skills, explore real use cases, and connect with other learners.

WHAT YOU'LL FIND INSIDE THE OPENAI ACADEMY
Once inside, you'll find that the Academy isn't a single course — it's a growing library of over 100 free tutorials and workshops, grouped into themed *"collections"*. Each collection focuses on how people in different roles are using AI in the real world.

Here's a quick snapshot of the types of courses and tutorials currently available through **OpenAI Academy** at time of writing. These examples show how the Academy helps people in different roles apply AI in real-world ways.

SAMPLE COURSES AT THE OPEN AI ACADEMY

Collection	Who It's For	What You'll Learn
ChatGPT at Work	Professionals and small-business owners	Practical ways to use ChatGPT for writing, research, data summaries, presentations, customer emails, and workflow automation.
OpenAI for Business	Managers and entrepreneurs	How companies are using AI to streamline operations, improve customer service, and automate repetitive tasks.
Developer Build Hours	Coders and tech-savvy users	Live and recorded sessions that show how to build apps and tools using the OpenAI API, GPTs, and plug-ins.
ChatGPT on Campus	Students and researchers	Tips for using AI ethically and effectively for studying, summarizing, research support, and academic writing.
Professors Teaching with OpenAI	Educators	Real classroom examples and teaching resources that show how universities are integrating AI into lessons.
OpenAI for Government	Public-sector professionals	Prompt packs and workflows for reports, policy writing, and public-service communication.
Sora Tutorials	Creators, marketers, and media professionals	How to generate, edit, and remix videos.

To join:

Go to: *academy.openai.com*

Sign in with your OpenAI account (or create one if you don't have one yet). Browse through course collections or topics that match your role. *e.g. business, education, development, etc.*

Enrol and start learning at your own pace.

Because it's self-paced and hands-on, you can begin with the basics (like prompt engineering or AI at work) and later move into advanced topics such as custom GPTs, tool integrations, or AI ethics. Over time, your prompt chops and workflow efficiency will grow far faster than if you were learning piecemeal from various blogs or random tutorials.

SUMMARY
New modules are added all the time, covering areas like AI ethics, creativity, productivity, and custom GPT development. The content is self-paced, short, and practical — most tutorials take less than 30 minutes and include real examples you can copy or adapt.

A FINAL TIP
Bookmark *academy.openai.com* and check back regularly — it's updated almost weekly with new tutorials and community build events. For anyone who wants to keep up with AI's rapid evolution, this is one of the best free learning hubs available.

GOOGLE SKILLS

The demand for AI expertise is at an all-time high, and as one of the global leaders in artificial intelligence, Google now offers nearly 3,000 courses, labs, and credentials under the banner *Google Skills.*

This new platform brings together learning resources from *Google Cloud, Google DeepMind, Grow with Google,* and *Google for Education* — all in one place.

Google Skills is designed for everyone — from students just getting started, to experienced developers seeking certification or business leaders exploring how to adopt AI effectively. Whatever your background, you can build and expand your skills on Google's comprehensive new learning hub.

WHAT YOU CAN LEARN
Learners can choose from a wide range of pathways. You might start with an entry-level *Google AI Essentials,* course from *Grow with Google,* then progress to more advanced programs such as *Google Cloud* certifications or *DeepMind's AI Research Foundations.*

If you're short on time, you can dip into *AI Boost Bites* — ten-minute lessons designed for quick, focused learning.

Hands-On, Practical Learning

Google Skills emphasizes *learning by doing.* You can earn skill badges, certificates, and full certifications, often through AI-powered labs that use *Gemini Code Assist* to guide you through real coding challenges. These experiences help transform theory into practical, job-ready skills.

SAMPLE FREE COURSES FROM GOOGLE SKILLS

Course Title	Duration	Platform/Format	Key Content/Audience
Introduction to Generative AI	45 mins to 1 hour	Google Cloud Skills Boost. Micro-learning with videos, reading, and a quiz.	**AI Fundamentals.** Explains what Generative AI is, how it works, and how it differs from traditional Machine Learning. It's a quick, **non-technical overview** for any beginner.
Introduction to Large Language Models (LLMs)	up to 1 hour	Google Cloud Skills Boost. Micro-learning with videos, reading, and a quiz.	**LLM Fundamentals.** Focuses specifically on Large Language Models—what they are (e.g., Transformers), common use cases, and how to use **prompt tuning** for better performance. **Non-technical.**
Machine Learning Crash Course (MLCC)	up to 15 hours	Google Developers site. Self-study with videos, text lessons, interactive visualizations, and coding exercises.	**Technical ML Basics.** A fast-paced, high-level introduction to core Machine Learning concepts (e.g., classification, deep learning). **Requires basic programming knowledge** (ideally Python) and is for a **technical dive.**

MOTIVATION THROUGH GAMIFICATION
Ninety-five percent of learners say they're more engaged with a gamified experience. Google Skills builds on that insight with tools such as *progress streaks, achievements,* and *shareable milestones* that make learning both social and rewarding.

GET STARTED — FREE OF CHARGE
Google Skills offers hundreds of no-cost AI and technology courses for all experience levels — no prior knowledge required. You can explore structured learning paths such as the *Generative AI Leader* series or simply browse topics that interest you.

Visit: *skills.google* to start learning today and future-proof your AI expertise.

THE AI FUTURE

The world is standing at the edge of one of the most transformative decades in history. Artificial Intelligence has already proven it can accelerate creativity, productivity, and innovation — but what comes next will make today's breakthroughs look modest.

WHERE WE'RE HEADED

Healthcare:
AI is moving beyond assistance and into true collaboration with medical professionals. Systems are already diagnosing cancers, predicting heart disease, and designing drugs faster than any research team could. Within the next decade, AI-driven *"digital doctors"* could manage much of our routine healthcare, freeing humans for complex, compassionate care.

Transport and Mobility:
Autonomous vehicles are not just about self-driving cars — they'll redefine logistics, shipping, and public transport. Expect cities where AI coordinates traffic lights, ride-sharing fleets, and energy grids in real time, cutting congestion and emissions.

AI as a Co-Creator:
We're entering an era of *augmented creativity*. Writers, filmmakers, musicians, and designers are already collaborating with AI tools to spark ideas, refine drafts, and build entire works in partnership. The boundary between human and machine imagination is becoming increasingly blurred.

AI and Sustainability:
Energy management, climate modelling, and smart agriculture are already being reshaped by AI. Intelligent systems can predict demand surges, optimize water use, and reduce waste

— critical steps in tackling climate change and feeding a growing population.

The Responsible AI Movement:
As power grows, so does responsibility. The conversation is shifting from *"what can AI do?"* to *"what should it do?"*. Ethics, transparency, and regulation will shape the global AI landscape — just as laws once shaped the early internet. The winners of this next era will be those who innovate *responsibly*.

Quantum Acceleration:
Quantum computing will supercharge AI's capabilities, turning today's *'impossible'* problems — from protein folding to global logistics — into solvable ones. A calculation that once took years, could be completed in minutes. This pairing, *Quantum + AI*, will redefine what's computationally possible.

Human + AI Partnerships:
The future isn't about replacement — it's about reinforcement. We'll move from using AI as a tool to treating it as a thinking partner. The most successful people and organizations will be those who know how to collaborate effectively with intelligent systems — asking better questions, interpreting better answers, and acting faster than ever before.

What It Means for You
The next few years won't be about who *uses* AI — nearly everyone will. The real divide will be between those who use it passively and those who use it *strategically*. The people who continually learn, experiment, and adapt will have a permanent edge.

AI won't eliminate creativity, intuition, or empathy — it will amplify them. It will make the curious more curious, the productive more productive, and the imaginative more imaginative.

Your challenge, and your opportunity, is to stay ahead of the curve. Keep exploring new tools. Keep refining your prompts. Keep questioning how AI fits into your world — not just how it changes it, but how it enhances what you do best.

LOOKING AHEAD
AI is still young — barely in its adolescence. But its trajectory is clear. Over the next decade we'll see:

- *Seamless voice-based AI companions that understand emotion and context.*

- *Autonomous systems managing homes, transport, and even city planning.*

- *Real-time translation that removes language barriers entirely.*

- *AI tutors, coaches, and creative partners available to everyone, everywhere.*

In short, a world that feels more connected, more efficient, and more intelligent — *provided we guide it wisely!*

A Personal Note
If you've reached this point, you're already ahead of most. You've not only learned how AI works, but how to *work with* it — and that's the mindset that will matter most in the years ahead.

The AI revolution isn't something happening *to* us. It's something happening *with* us. So, keep learning. Keep experimenting. And keep imagining what's possible.

The future belongs to the curious.

Peter Thorpe
Author

AND FINALLY...

I think it's only fitting that ChatGPT should have the last word.

ME: *"Thank you, ChatGPT, for all the help you've given me in writing this book. I couldn't have done it without you!"*

CHATGPT: *"You're very welcome, Peter! It's been a pleasure to help bring The AI Future to life. Remember — the best way to stay ahead is to keep learning, experimenting, and asking great questions.*

I'll be here whenever you need a hand, an idea, or a fresh perspective.

Happy writing — and here's to your AI-powered future!"